The Arctic Skua

*a study of the ecology and evolution
of a seabird*

The Arctic Skua

a study of the ecology and evolution of a seabird

PETER O'DONALD, Sc.D

Fellow and Tutor, Emmanuel College, Cambridge

ILLUSTRATED BY ROBERT GILLMOR

CAMBRIDGE UNIVERSITY PRESS

CAMBRIDGE

LONDON   NEW YORK   NEW ROCHELLE

MELBOURNE   SYDNEY

Published by the Press Syndicate of the University of Cambridge
The Pitt Building, Trumpington Street, Cambridge CB2 1RP
32 East 57th Street, New York, NY 10022, USA
296 Beaconsfield Parade, Middle Park, Melbourne 3206, Australia

First published 1983

Printed in Great Britain at the University Press, Cambridge

Library of Congress catalogue card number: 82-12782

British Library Cataloguing in Publication Data

O'Donald, Peter
The Arctic Skua.
1. Stercorarius parasiticus
I. Title
598'.33   QL696.C488 *636*   *1983*

ISBN 0 521 23581 2

39,833

AL

Contents

v

List of figures

Figures 7.1, 7.2, 7.3 and 7.4 were drawn by Robert Gillmor on
 the basis of photographs and drawings in a paper by Perdeck
 (1963)

List of tables

Preface

This book describes the results of research spanning a period of 30 years. The late Kenneth Williamson began studying the Arctic Skuas of Fair Isle in 1948. Only 15 pairs were then nesting on the island. In 1957, Peter Davis took up the study, having succeeded Williamson as Warden of the Fair Isle Bird Observatory. As part of research for the Ph.D. degree of the University of Cambridge, I spent three years from 1958 to 1961 studying the genetics of the Arctic Skua. I was supported by a Nature Conservancy Research Studentship. This initial period of research on the Arctic Skuas ended in 1962 when Peter Davis left Fair Isle. By this time, 71 pairs were nesting on the island.

In 1970, R. J. Berry and Peter Davis published a paper analysing the breeding dates of the different phenotypes of the Arctic Skuas (Berry & Davis, 1970). They found that pale males, breeding for the first time, bred several days later on average than first-time, intermediate and dark males. They interpreted this as a behavioural adaptation of pale birds to their more northerly distribution where later breeding might be advantageous. At that time, I was working on models of Darwin's theory of sexual selection. The later breeding of certain male phenotypes in new pairs is exactly what the models predict, whereas the adaptation postulated by Berry and Davis should have been shown by all birds. This was obviously an opportunity to test Darwin's theory in a natural population. I successfully applied for a Research Grant from the Natural Environment Research Council (NERC) for the support of three years research, 1973–75. I later obtained another grant from NERC for the three years, 1976–79. In the period 1973–75, the grant allowed for the appointment of John F. Davis as a full-time research assistant. This second period of research on the Arctic Skuas of Fair Isle ended in 1979.

When John Davis and I began the second period of research in 1973, 106 pairs of Arctic Skuas were nesting on Fair Isle. John Davis devised the method of catching the adult breeding birds in funnel traps. This was much quicker than the clap nets previously used. Most of the breeding birds were caught and ringed in 1973. I am very pleased to acknowledge the great enthusiasm and energy which John Davis put into studying the Arctic Skuas. In 1975, Shoshana Ashkenazi assisted in our survey of the Arctic Skuas of Foula. Roger Broad, the new Warden of the Bird Observatory, gave much of his time to the skua study, particularly in helping to catch the birds. I am grateful for his very considerable help in 1976 after the first NERC grant had expired. Later, in 1978 and 1979, I was assisted by Jane French, a research student on an NERC Studentship, and Iain Robertson, who had succeeded Roger Broad as Warden of Fair Isle.

This book is a largely original account of the research on breeding behaviour, ecology and evolutionary genetics of the Arctic Skua. I have added a chapter on feeding behaviour (chapter 3) based on the published work of others, though with my own analyses of the data. Section 7.1 (chapter 7) on mating behaviour closely follows Perdeck's descriptions of nesting, courtship and copulation (Perdeck, 1963). Chapter 2 on numbers and distribution is a synthesis from many sources. The rest of the book is original. It describes the results of analyses of data from Fair Isle. I have already published a number of papers on the genetics, demography and sexual selection of the Arctic Skua analysing data for the period up to 1976. In this book, the data are now complete to the end of the breeding season in 1979. Chapters 9 and 10 are based on two published papers (O'Donald, 1980*b*,*c*), but extended with new theory and analyses. The data of chapter 4 on breeding ecology have never previously been published in any form. Some of our earlier conclusions on differences between pale and melanic males in the sizes of their territories (Davis & O'Donald, 1976*b*) have been contradicted by the subsequent data. In sections 7.3 and 7.4 (chapter 7), I analyse the complete data on territory size, finding no phenotypic differences, but a slight, statistically significant relationship of breeding date to territory size. In 1978 and 1979, my brother-in-law, Terry Lynch, assisted in the mapping of the nests and territories. He also took many photographs of Arctic Skuas, including those of the nest with eggs and the pale, intermediate and dark phenotypes of chicks which form figures 1.2, 1.4, 1.5 and 1.6. These are prints from colour slides.

In acknowledgement, I am most indebted to the Nature Conservancy, later the Natural Environment Research Council, first for the Research Studentship with which I began my career of research in population genetics, and then for the two Research Grants which supported the study of the Arctic Skuas in the period 1973–79. To all those who have helped me at various times – Peter Davis, John Davis, Roger Broad, Shoshana Ashkenazi, Jane French, Terry Lynch and Iain Robertson – I offer my sincerest thanks. Finally, I thank Miss Rena Beech who typed this book from my manuscript.

1

The Arctic Skuas of Fair Isle

1.1 The Arctic Skua and its breeding grounds

The Arctic Skua is a circumpolar Arctic species that breeds as far north as Spitzbergen. At the southern end of its range, it breeds in small colonies in parts of the Hebrides and the north of Scotland. But in the British Isles it is common only in Shetland (59°N) where it is fairly widespread. Large colonies are found on the islands of Fair Isle, Foula, Bressay, Noss, Fetlar and Unst. Isolated pairs and loose colonies can be found throughout Shetland. Recently it has been giving ground to the Great Skua or Bonxie. In 1950, the largest Shetland colony of Arctic Skuas was to be found in the bird reserve of Hermaness on Unst, the most northerly point of the British Isles. This colony was completely overwhelmed by Bonxies as they rapidly increased in numbers between 1950 and 1975, completely taking over the bird reserve. The largest compact colonies of Arctic Skuas are now to be found on Fair Isle and Foula: about 130 pairs breed on Fair Isle and 300 pairs on Foula. Both colonies are under attack. The Foula colony is being pushed into a smaller and smaller area by the spread of the Bonxies. On Fair Isle, the predators are human. Unfortunately, Arctic Skuas are recorded as having been shot on Fair Isle – bodies with bullet holes having been found. In spite of such depredations about 120–140 pairs have bred since 1975.

The Arctic Skuas of Shetland on the islands of Fair Isle and Foula are the subjects of this book. I studied them first in the period 1958–61, when I was a research student collaborating with Peter Davis who was then Warden of the Fair Isle Bird Observatory. Twelve years later, a Lecturer at Cambridge University, I returned to Fair Isle to test theories of sexual selection in the Arctic Skuas having obtained a research grant from the Natural Environment Research Council. I had set up and analysed mathematical models of Darwin's theories of sexual selection in birds and proposed to test these models with data of the matings and relative fitnesses of the Arctic Skuas. This research came to an end in 1979. By that time I had collected enough demographic data on the Arctic Skuas to obtain good estimates of the parameters of sexual selection. Using statistical tests, I could refute specific models of the mechanism of selection. At the beginning of this second period of research I was most fortunate to have the assistance of John W.F. Davis. He spent four years working with me on the project from 1973–76. In the later years, 1977–79, Roger Broad and Iain Robertson, Wardens of the Bird Observatory, generously assisted in catching the adult birds for ringing.

The season of field work on the Arctic Skuas of Shetland begins in mid-May when the first pairs lay their eggs. The first eggs are usually laid on the same day each year. The birds are remarkably precise and consistent in timing when they breed. My own arrival in Shetland was seldom timed as precisely as the birds'. British Airways' flights to Shetland are often delayed by fog, sometimes for several days, for Sumburgh Airport is built on a narrow isthmus of land at the southern tip of Shetland where sea fog is very likely to occur in summer. So relief was often mixed with pleasure as the plane at last set down on the runway at Sumburgh. If British Airways could get in to Sumburgh, it was certain that Loganair could take one of their eight-seat Islander aircraft to Fair Isle, to land 200 ft up on the gravel airstrip situated in the middle of Fair Isle in the Arctic Skua colony.

Until 1973, only a rough stony track constructed during the war existed to justify the name 'airstrip'. From the area round this original airstrip, the Arctic Skuas began their colonization of Fair Isle. Their original nests were found on either side of the road to Ward Hill where it crossed the airstrip. From this centre, the Arctic Skuas quickly spread to Sukka Mire, Byerwall and Homisdale and up onto the heather moorland of Swey. They crossed Homisdale to Eas Brecks. The colony's centre of gravity gradually moved north. By 1977, Wirvie Brecks had become the most densely colonized area. It is easy to see why a colony should move its centre like this. Young birds cannot displace the holders of the original territories. They establish new territories for themselves at edges of the colony. As they become older they give rise to a new centre from which the colony can expand in new directions. The old centre becomes less populous as the original birds die. Some new birds may now establish themselves in territories where birds have died, but many new birds look for areas where there are no existing pairs to compete with. The maps in figure 1.1 show how the colony has expanded and moved since 1948.

By the time the modern airstrip was under construction in 1973, the airstrip area had already become less densely occupied. The construction work itself, however, caused little disturbance. Pairs continued incubating only 20–30 m away from the continual movement and noise of the bulldozers. This is one of the great advantages of the Arctic Skua as a bird to study: it seems almost impossible to put the birds off their nests. Regular checks of nests to determine the laying and hatching dates of the eggs and

Figure 1.1 Map of Fair Isle showing the limits of the Arctic Skua colony in different years. Nests of isolated pairs in 1976 are shown as clear white circles.

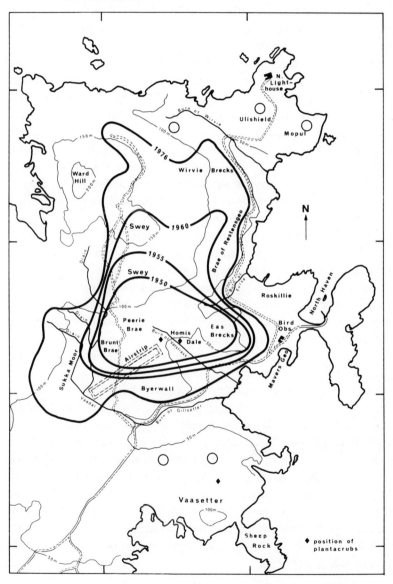

the fledging of chicks have no apparent effect at all. After the eggs in a nest have been checked, for example, the birds are usually back at the nest incubating again within a few minutes. Trapping birds on their nests while they are incubating the eggs – as we shall see, this is the only practical way of catching them – never seems to upset them or put them off: they are back sitting on the eggs as soon as they have been ringed, released and the traps removed.

Some pairs do still nest near the airstrip. East of the airstrip, on Byerwall, nests are densely packed, and here also 'clubs' of young, first and second year birds often congregate round a small water hole. Air traffic is only a momentary disturbance, even to birds nesting by the side of the runway. As the map shows, the airstrip lies in a SW–NE direction on the moorland above Homisdale. Unless the wind is blowing strongly from the north or east, the plane lands towards the SW, approaching over Eas Brecks, which is another of the more densely colonized areas. The new Bird Observatory is situated just below Eas Brecks on the road to North Haven. From the airstrip, a direct line to the Bird Observatory takes you across part of Byerwall, down into Homisdale, up over Eas Brecks and down again to the Observatory. These areas all form parts of the Arctic Skuas' colony. They show the different habitats that Arctic Skuas choose for nesting.

Byerwall, like much of the higher ground of Fair Isle, consists of short tufts of heather (*Calluna vulgaris*) with some Crowberry (*Empetrum nigrum*) separated by stony patches of ground, moss, *Nardus stricta*, Fescue and other grasses. Byerwall dries rapidly after heavy rain. The rocky subsoil is only two or three inches below the surface and surface water drains quickly away. This barren aspect of the higher ground was produced by peat cutting. Fair Isle was overlain with peat, as much of Shetland still is today. Most of the peat was cut in the nineteenth century, when the population rose to an almost insupportable 350 people. Some peat is still left on the west side of the island below Ward Hill. Peat cutting continues there today. When a peat bank has been cut, the top few inches of turf are relaid over the rocky subsoil where the peat was removed. Thus only a very shallow layer may be left on top of the subsoil. Arctic Skuas nest among the heather and rocks of such areas as Byerwall, Eas Brecks, Swey and Wirvie Brecks where the original peat has been almost completely removed. Except on the steep slopes below Swey, the heather is very short and stunted: sheep graze any new growth that appears. Over-grazing probably explains why these areas remain so barren and rocky. But this seems no problem to the skuas. A shallow scrape in the grass or heather is a sufficient nest for them (Figure 1.2).

On Byerwall, which is a fairly densely populated part of the colony, the nests are about 80 m apart. The nesting area is quite level and shielded by rising ground from the road to the runway and the hut where passengers wait for the plane. On climbing over this rise and looking down on Byerwall, birds will be seen flying off their nests. Many nests can be found

Figure 1.2 An Arctic Skua's nest with eggs.

quite easily, simply by walking through the colony and trying to note the spot where a bird was seen to fly up from. To approach over a rise is the best way of spotting them fly up. Unlike a bird merely sitting in the heather, an incubating bird gets up immediately and flies straight into the air. A bird sitting in the heather will probably stand up and look round for a few moments before taking flight. By remaining alert to notice the spot where a bird suddenly came up from, you can often walk straight to the nest. As you approach, the pair of birds fly over their territory in a slow, undulating flight making anxious, low, squeaking calls. Other nearby pairs will attack the threatened pair if they cross the territorial boundaries. The intruding ornithologist may then be forgotten for a few moments while a furious aerial chase takes place. The birds scream round the sky in tight turns making their wild squealing cries *eeeow eeow*. Back over their own territory, the chase over, they will often attack as you approach. When you are near the nest itself, they will flop out of the sky in their distraction behaviour. They always start their attempts at distraction in the same way. They suddenly seem to stop flying and flutter to the ground. The distraction display itself varies greatly between individual birds. Some birds go almost frantic: they jump up and down squeaking loudly, helplessly beating their wings on the ground. Some scurry about, wings spread and drooped, in the characteristic injury-feigning pattern of behaviour. Other birds are simply cowards, running away to watch from a distance and doing nothing. These cowards are really the most annoying although they never press home an attack: they are off the nest at the first hint of danger and keep well away whatever happens. It is sometimes almost impossible to get a good view of them, and particularly difficult to get a good view of the combination of colour rings which uniquely identifies each individual bird.

The bravest birds are usually the easiest to identify: they yell and scream and injury-feign just a few metres from you. Their combination of colour rings is easy to see as they jump up and down. This behaviour usually lasts for only a few minutes. After brief attempts at distraction, the birds then attack. Those with the most vigorous distraction behaviour are often the most aggressive. They drive you from their territory, diving at you at high speed. They often strike with extended feet. Robert Gillmor's drawing shows an attacking skua about to strike. Walking through the colony without protection often brings many painful buffets to the head. We usually carry a cane held above our heads for the birds to aim at and hit.

The nests are marked with canes, placed a few metres away from the nest itself. A tag on the top of the cane gives the number of the pair. Viewed from the rising ground above Byerwall, canes can be seen dotted here and

A dark Arctic Skua flies up from its nest on Eas Brecks, Sheep Craig in the background.

An Arctic Skua attacks. Its mate feigns injury on the ground.

there. Some birds immediately fly off their nests; others can be seen through binoculars sitting on their nests beside the canes. They soon get used to the daily checking of each nest. Often they will incubate until you are no more than about 100 m away. They settle back on their nests as soon as you have passed through their territories.

From the dry heather moor of Byerwall the ground dips sharply into Homisdale. This is one of the most sheltered parts of Fair Isle. The valley floor is a sphagnum bog through which the Burn of Vatstrass flows in a straight drainage channel. In May and June, when the skuas are nesting, it is colourful with lady's smocks (*Cardamine pratensis*), cotton grasses (*Eriophorum augustifolium*), butterworts (*Pinguicula vulgaris*) and spotted orchids (*Dactylorhiza fuchsii*). It becomes drier towards the head of the valley. Here it is pasture – a small area of excellent grazing – before giving way to heather on the upper slopes. The plantacrubs marked on the map are an interesting agricultural survival. They are small stone-built enclosures in which seedlings can be grown protected from the wind. Most of them are now ruined, but one is still occasionally used to rear seedling cabbages. They are useful places to hide in while looking for the skuas' nests in Homisdale or waiting for a bird to enter a trap. The nesting habitat of Homisdale is quite different from Byerwall. The nest is often to be found in the middle of a tuft of grass raised slightly above the level of the bog. The nest scrape may be quite damp. This would presumably chill the eggs more rapidly at those times when the birds are not incubating. But nests in these boggy places suffer no significant reduction in hatching or fledging rates.

Continuing northwards towards the Bird Observatory, the ground rises gently to the heather moor of Eas Brecks. This is a densely colonized area similar in habitat to Byerwall. To the west it dips into the shallow valley carrying the Burn of Furse northwards from the head of Homisdale to the sea. The skuas have colonized the whole of the area of Furse and Eas Brecks lying south of the road to the North Lighthouse. Above Furse the ground rises steeply through deep heather to the moorland of Swey, about 4–500 ft (112–153 m) above sea level. The skuas seem to be unable to nest on these steep slopes. But eider ducks make their warm, down-lined nests under the protection of the heather bushes. When the plateau of Swey has been reached, the skuas again occupy the ground. This is a rocky, stony area like Byerwall and Eas Brecks. In a few places, narrow peat banks have been left uncut. The ground is thus uneven, with raised ledges of peat. But it is just as barren as Byerwall and Eas Brecks – more so perhaps – with larger rocky outcrops. This continues northwards from Swey. The ground slopes gently down across a mile of moorland towards the road to the

North Lighthouse. Skuas occupy the whole area to the west of the road and south from Wirvie Brecks. Wirvie Brecks is the most northerly area yet to have been colonized by the Arctic Skuas. It is now the densest part of the whole colony. The nests are only about 40 m apart in some places. Further west, the Bonxies are in occupation. They set up their territories earlier than the Arctic Skuas who are then unable to dislodge them from ground Arctic Skuas formerly held. Thus, in competition, the Bonxies always win the territories. They have indeed expanded rapidly throughout Shetland, sometimes at the expense of colonies of Arctic Skuas. But for their nesting sites, Bonxies prefer wetter, less rocky areas: they always make their nests in a grassy spot, never in the heather.

South of Wirvie Brecks, the Arctic Skua colony continues along the Brae of Restensgeo joining up with the parts of the colony on Swey and Eas Brecks. Most recently, in 1978 and 1979, two pairs have nested on the heather at Roskillie on the other side of the road to Eas Brecks. The road itself continues round the east side of Eas Brecks to reach the Bird Observatory in a green 'cup' overlooking Mavers Geo and the North Haven.

Any brief tour of an Arctic Skua colony will show that it is polymorphic. Most birds are a dark brown, but about a quarter are pale with a white belly, breast and neck. Some of the pale birds have a darker band across the breast between neck and belly; others have a completely white front. A closer look at the dark birds will show that they also vary between themselves. Some have a collar of paler feathers round the neck. This may vary from an almost white neck to an only slightly lighter brown neck. The belly may also appear somewhat paler or flecked in those birds with pale or white collars. Other birds are completely dark brown, with no paler neck or belly feathers. This is a striking example of a polymorphism in plumage. The different types of plumage are called the phenotypes. We have distinguished three main phenotypes: pale, non-melanic birds with white belly and neck; intermediate melanic birds with brown or dark brown belly and paler or white collars round the neck; and dark melanic birds, uniformly dark brown. These phenotypes have increasing amounts of dark brown melanic pigment deposited in the feathers. Somewhat arbitrarily, we may separate the intermediates into two classes: intermediates with some white neck feathers and slightly lighter brown belly feathers, and dark intermediates recognizable as having only a few pale straw-coloured or yellow feathers on the sides of the neck and ear coverts giving the neck a slightly paler look beneath a cap of dark brown feathers on the crown.

Matings can be seen between any of these phenotypes. Most pairs are matings of intermediates with intermediates or intermediates with either darks or pales, for most birds are intermediates. I first saw this polymorphism in 1958 when I was spending a holiday on Fair Isle. It seemed an obvious subject for genetic and ecological study. K. Williamson, the first Warden of the Bird Observatory, had already begun an intensive field study of the Fair Isle Arctic Skuas (see Williamson, 1965). P.E. Davis, who succeeded as Warden, and myself, then a research student, carried this work on until 1962. I resumed the work in 1973 with a Research Grant from the Natural Environment Research Council.

Any sharp differences in morphology, like the different phenotypes of the Arctic Skuas, are almost certainly genetic. More continuous phenotypic variation may be expected to be determined by both environmental and genetic factors. The polymorphism is widespread (Southern, 1943); it is stable if we can judge from counts of pale and dark birds made by a number of ornithologists over many years (Jackson, 1966; O'Donald & Davis, 1959; O'Donald, 1980c; Pennie, 1953; Perry, 1948; Yeates, 1948); and it is clinal, changing from a high frequency of melanics at the southern end of the Arctic Skua's range to about 100 percent pale at the northern end. A stable clinal polymorphism of strikingly different phenotypes clearly invites ecological and genetic study: balanced selection, or selection and migration, almost certainly maintains the stability of the polymorphism at each point; ecological variation along the cline gives rise to changes of selective value at different points; the adaptive significance of the polymorphism can thus be analysed in relation to ecological factors. By ringing the adult birds, their survival and reproductive rates can be measured throughout their adult lives. If all the nests can be found and all the breeding adults ringed, complete demographic data can be obtained on all the birds in the population. Thus the selection acting on the different phenotypes can be measured. In this respect, birds are often better organisms to study than animals of many other groups: nests are easily found and inspected; the adults can usually be caught. The whole population can be marked and identified and individuals' fates followed throughout their lives. With their exposed nests on easily accessible moorland and bog, the Arctic Skuas seemed almost ideal for detailed ecological and genetic research. The next section describes the methods we used to study them.

1.2 Catching the Arctic Skuas

Arctic Skuas start to return to their Shetland breeding grounds in April. On Fair Isle, the first birds arrive between 16 and 18 April. By the end of April perhaps about one third of the adults may have returned. But it is difficult to determine the numbers accurately until eggs have been laid and separate pairs can be identified. The first eggs appear between 16 and 18 May. The first eggs hatch between 11 and 13 June, usually on 12 June. For the period 1973–79, pair 106 has consistently been the first to hatch their eggs, the first chick almost always appearing on 12 June. They thus incubate for about 26 days. The extraordinary consistency of the dates when particular pairs breed strongly implies that the timing of reproduction is under precise photoperiodic control. But, as we shall see, this consistency is shown only by pairs who have bred together in previous years: either they mutually adjust the timing of their return to the island and the date when they breed, or one of the pair, probably the male, returns first and the female determines the date of mating and reproduction.

The adult birds can only be caught at the nest. We have experimented with a compressed air gun firing a net over a spot where birds tend to congregate – near a pond for example – but they so seldom venture into range that we never caught any by this method. Near the nest we were occasionally successful catching birds in a fine 'mist net' extended on high poles behind a stuffed Bonxie. A stuffed Bonxie, if placed beside the nest, produces a violent aggressive reaction. We caught a few birds in the mist net as they dived at the stuffed Bonxie. But apart from this, a trap to catch a bird sitting on its eggs seems to be the only reliable method. Almost all the adult breeding birds can be caught in this way. Originally, the birds were

A pale Arctic Skua prepares to land by its nest.

caught in 'clap nets' worked from a nearby hide. This is a net supported by a cane which can swivel on a loop of wire pegged into the ground. A string is is attached to the top of the cane and leads at an acute angle to a hide almost 30 m away. A sharp pull on the string brings the cane up over centre so that the net falls across the nest. The real eggs are removed to prevent their being smashed when the bird is caught and flaps up and down in the net. They are wrapped in cotton wool and placed in a tin to keep them warm; dummy eggs are placed in the nest. This method of trapping is satisfactory but very slow. Some birds are naturally reluctant to sit when the net is in position. About 40 min is the maximum time we have been prepared to wait for a bird to come back on the nest: in our experience they seldom come back after this length of time; the eggs will then have started to chill and there is also a danger of putting the pair off the nest completely. After a while, a bird seems to lose its broodiness and refuses to sit. If the other bird of the pair were also disinclined to brood, the eggs would then become fatally chilled. Fortunately very few eggs have failed to hatch. Addled eggs are found only rarely and do not appear to be the result of catching the birds.

Sometimes, if we were lucky in trapping, both birds of a pair were caught in rapid succession – one perhaps only five minutes after the other. But more usually only one of the birds is feeling broody, the other making no attempt to come back to the nest. Since both sexes share incubation, one bird presumably starts to become broody after the other has sat for a given length of time. The sitting bird loses its inclination to brood and leaves the nest to its mate who has become broody in the meantime. Only one bird would normally be broody at a time – the one incubating the eggs. Towards the end of a period of incubation both might be broody and both would then be prepared to sit and be caught. But often we would have to make several attempts over a period of many days to catch a particular bird that seemed reluctant to brood. One or two birds, indeed, we were never able to catch: they either left everything to their partner, or were just not prepared to approach a trap.

Setting up the nets and hides and waiting for the birds to return obviously take time and effort. When Dr J.W.F. Davis started working with me on the project in 1973, he introduced the method of using funnel traps to catch the birds. These are traps of galvanized wire netting about 60 cm long by 40 cm wide by 30 cm high, with a funnel-shaped entrance. A funnel trap is drawn in figure 1.3. The traps are staked down over the nests. It is quite important how they are placed. Often a bird may have a particular approach to the nest. The funnel must be placed to face the way the bird normally approaches. On a windy day, birds generally land into

wind towards the nest. The funnel must then face down wind so that it can
be approached into wind. Dummy eggs are, of course, placed in the nest
and the traps watched from a distance. Once the bird has been seen to enter
the trap and sit on the eggs, the ornithologists charge towards the trap from
the funnel side, driving the bird to the back of the trap. Usually the bird
flaps about at the back of the trap but some do manage to escape. An escape
is always particularly unfortunate, for a bird who has once been in a trap
will obviously be reluctant to enter one again. Generally, however, these
traps cause little disturbance in the colony: no hide is necessary and they are
set up in two or three minutes. Several traps can be used simultaneously.
We often set up three traps in three different parts of the colony. Each trap
could then be observed at regular intervals through binoculars. With luck
three birds might thus be caught in half an hour. By using these traps we
could ring almost all the adult birds in the colony. The colony had
expanded from 67 pairs in 1961 to 106 pairs in 1973. In 1973, John Davis
and I were able to catch all but 24 of the 212 breeding birds. Most of the
uncaught birds were late breeders, probably new pairs. In the subsequent
years it was only necessary to catch the unringed birds, most of which were
new immigrants to the breeding colony. At the end of the 1974 season, only
ten birds remained unringed. Of course fewer birds had to be caught – the
unringed birds from the previous year and the new immigrants. But some
birds always refused to enter the traps. We were usually able to catch these
by the old method of clap netting. The odd one or two proved to be
uncatchable, even when days were spent gradually accustoming them to the
presence of the clap nets.

Birds that remain unringed one year may produce ambiguity next year

Figure 1.3 A funnel trap for catching Arctic Skuas.

whether they are new to the colony or the same bird that bred in the previous year. To study natural and sexual selection, we need to know the survival rates of the different phenotypes and their reproductive success. Part of the variation in reproductive success may depend on whether the birds have changed their mate and hence on their relative success in finding a new mate. This is where sexual selection may operate. A bird who can find a new mate quickly will be more successful than one which takes longer: the quicker bird will produce its eggs earlier in the season; it will then rear more offspring because, exactly as Darwin (1871) originally predicted, early pairs are more successful than later ones (see chapter 8, section 8.3 for a detailed discussion of Darwin's theory of sexual selection in monogamous birds). To study the ecological genetics of the Arctic Skua, we must therefore obtain data of survival, change of mate and reproductive rate of each of the different Arctic Skua phenotypes.

From these data, selection can be analysed into its natural and sexual components. New pairs supply data on how change of mate affects reproductive success, hence data on sexual selection. It was therefore important to identify as many birds as possible from one year to the next and thus ascertain if they had changed mates. Proceeding by elimination of known pairs from previous years and new pairs known from ringing, only a very few pairs could not be identified definitely as either new or old. In the years after 1974, almost all birds were ringed and ambiguities did not arise. Very little data was in fact lost because some pairs could not be identified. Often the only birds not to be caught were very late breeders, presumably new pairs, that failed to hatch an egg or rear a chick. By 1977, when the colony consisted of 139 pairs, only four adults were still unringed at the end of the breeding season.

1.3 Studying the birds and recording their data

For a combined ecological and genetic study of a polymorphic population, we need to obtain demographic data of the survival and reproduction of the genotypes that determine the polymorphism. Ideally we should obtain data of the survival and reproduction of cohorts of each genotype. We should follow the fortunes of a group of newly fledged chicks. If chicks always returned to their natal colony to breed this could be done by ringing all the chicks. The chicks that survived to breed would then be identified when the breeding adults were caught. Since Arctic Skuas generally breed for the first time at the age of four or five years, a very long-term programme of field work would be necessary. But unfortunately a proportion of chicks migrate to breed in colonies other than the one where they were born. Chicks ringed on other islands have been caught as breeding adults on Fair Isle. A proportion also lose their rings. Most of the new breeding adults coming to Fair Isle up to 1977 were in fact unringed and their origin unknown. Most of these unringed birds may well have been born on Fair Isle before our project had begun again in 1973. They would not have been ringed as chicks. In 1977 many of the new adults could be identified as Fair Isle chicks returning to breed. A few birds ringed as chicks on other islands have come to breed on Fair Isle, but since only a small proportion of all Shetland Arctic Skua chicks are ringed in a season we cannot say what the proportion of migrants is likely to be. In the previous period up to 1961 when Arctic Skuas were studied on Fair Isle, very few Fair Isle chicks were recovered as breeding adults. But the rings used then were of soft aluminium that quickly wore away. They dropped off after a few years. Today's rings are much harder and show hardly any wear after five years. A strict cohort analysis is clearly impossible since it is not known what proportion may have migrated elsewhere. Even if it were possible, a cohort analysis would be impractical. A colony of 100 pairs may produce 120 chicks in a year. Of these, only a small proportion will survive to breed as adults. Yet we would be trying to detect differences in survival and reproductive rates between individuals of different genotypes. We would have to follow successive cohorts over very many years to get samples large enough for this purpose. All we can do is to follow the survival and reproduction of adults from the time they first breed on Fair Isle. Of course, we shall identify a number of Fair Isle chicks who have returned to breed. These can supply data on the age when Arctic Skuas first breed and the genetics of the phenotypes, but their number can hardly be sufficient to detect relative differences in survival rate: the probability of a chick's survival to breed cannot be estimated accurately enough.

Improvements in technology have greatly increased the efficiency of the study of birds by ringing. Plastic colour rings often used to drop off between seasons, many birds having to be re-caught or their identities guessed at. Now we use tough plastic rings glued by a powerful adhesive. Ring losses are minimal. A bird once caught need hardly ever be caught again: it can always be identified by its colour rings. Fewer than ten birds have had to be re-caught because of the loss of a colour ring. Of course, to identify a bird a colour combination chart must be used. For identification, we have used three colour rings in combination with the birds' phenotypes and their numbered metal rings. A bird is known by its ring number. Each bird has a card kept in a card index in ring number sequence. On the card is recorded the colour ring combination, the bird's phenotype and sex and the numbers of the nests where it mated each year. Each pair is given a nest number which is retained as long as the two original birds stay together. For example, pair 106 stayed together from 1973, when they were first ringed, until 1979. So these birds' cards record the fact that they bred together at nest 106 in the years 1973 to 1979.

Two rings are placed on each leg. On one leg is a numbered metal ring of the British Trust for Ornithology (BTO). This identifies the bird by ring number. Above it is placed a coloured plastic ring. Two coloured plastic rings are placed on the other leg. We use the following colours for coloured rings placed above the BTO ring:

> Orange (O), Blue (Bl), Light Green (Lg), Brown (Bn), White (W), Yellow (Y), and Grey (Gy).

For the other leg we use the colours:

> Blue, Red (R), Green (G), Yellow, and White.

Thus a dark bird might be described as being

> Bl/W left; Bl/BTO right

This means its rings are blue on top of white on the left leg, and blue on top of BTO on the right leg. Reference to the colour ring combination chart shows that a dark bird with this combination must be number EF.84543, first caught and ringed at nest 118 in 1973 and which continued to breed there up to and including 1979. As each bird is identified, its ring number card is moved forward from the sequence of cards of birds in previous years to the sequence of cards of birds identified as present in the colony in the current year. The card for a bird which fails to return to the colony is left among the cards for the year in which it was last identified. The card index of ring numbers thus provides the data of the survival of birds from one year to the next. As new birds are ringed their cards are placed in ring number sequence with the cards for the birds of the current year.

The birds can be identified once they start defending their territories and showing distraction behaviour. They usually feign injury close to an intruder. Their colour ring combination can then be noted. (But the rings of the timid birds who retreat to leave their nest undefended, may be exasperatingly difficult to see.) Having marked the position of the nest with a cane and checked the colour combinations of the birds, their ring numbers are found on return to the observatory from the colour combination chart. Their cards are updated. If they are the same pair that bred together in the previous year, they are given their old nest number, which is recorded on their cards. If they are birds which have not previously bred together, they are given a new nest number from a higher sequence.

Pairs with one or both birds unringed are almost always new pairs. Some birds, of course, elude capture and are left unringed at the end of a season. But ambiguity will arise only if the unringed bird has the same mate, is of the same phenotype and the pair is on the same territory as the pair containing the corresponding unringed bird from the previous year. These cases of ambiguity are very rare, for at least 95 per cent of the breeding adults have been colour ringed since 1974.

Metal tags bearing the nest numbers are attached to the canes marking the nests. Nest record cards are made out every year for every nest. Details of each pair's breeding are entered on their nest record card. Ring numbers and colour ring combinations are also entered on the nest record cards, which are thus cross-referenced to the ring number cards of individual birds. The breeding data are obtained by regularly checking the nests of all pairs. We make nest checks every day or every second day depending on the weather. Laying dates are recorded if known. But it may not be possible to say when the first egg was laid, for nests can only be found when they contain eggs. If a nest is found to contain only one egg, the date when the second egg is laid will be ascertained during subsequent nest checks. Usually, however, nests are found with their full complement of two eggs. Some pairs have regularly produced three eggs, and clutches of four eggs have been recorded in the literature (Witherby, Jourdain, Ticehurst & Tucker, 1941). But eggs in clutches of three have never hatched.

The important data obtained from the nest checks are (i) number of eggs in the clutch – usually one or two; (ii) the date when the first egg hatched; (iii) the number of chicks that successfully hatched; (iv) the number of chicks that successfully fledged; and (v) the ring numbers, and phenotypes of the fledglings. These data are all entered on the nest record cards against the date when the observations were made. The hatching date is used as a measure of breeding date. In new pairs, this will be determined partly by the time taken to find a new mate. Old pairs generally breed about nine days earlier in the season than new pairs. This difference reflects the time required to find and court a new mate. If phenotypes differ in their attractiveness or ability to attract or compete for mates, they will differ in breeding date. This will give rise to sexual selection. Differences between the phenotypes in age of maturity, survival rate and reproductive rate can be used to estimate the components of natural and sexual selection.

To detect the effects of sexual selection, the birds must of course be sexed. But the sexes are phenotypically identical in plumage and cannot be separated on morphological characters. Some birds are sexed simply by observing them copulate while we walk through the colony. Most birds can be sexed by examination when they have been caught for ringing. The female, having laid the eggs, has the larger cloaca. This will be particularly obvious if an egg has recently been laid. The relative sizes of the cloacas of a pair of birds has been a very reliable method of determining sex. But the cloacal difference gradually disappears after the eggs have been laid. After two or three weeks the difference between male and female may have become too small to be useful for sex determination. Fortunately, changes of mate can be used to sex many birds. Often the sex of a large number of birds can be ascertained from the previous matings of a bird whose own sex was ascertained when it mated with a bird of known sex. Thus, each year, new pairings of known birds are used to determine the sex of their new mates whose sex then determines the sex of their previous mates, and so on. About 95 per cent of all adults have thus eventually been sexed. Birds found dead or shot or killed can of course be sexed positively by dissection of their sex organs. Unhappily a number of birds appear to have been shot over the years, but at least, their sex had always been correctly ascertained.

The Arctic Skuas are strongly territorial. We have attempted to map their nests and territories. The territories can be mapped by fitting contiguous polygons round their nests. Chapter 7 describes the method and the assumptions it is based on. It depends on first mapping the nests. This has been done by Brunton compass and measuring tape, which gives a very accurate map of the positions of the nests. Before the eggs have hatched, the canes are placed about 10 m from the nests in particular directions relative to outstanding features. The nests can easily be found if their direction from the cane is known. If the canes had been placed close to the nests, it would have been too easy for some people to find them and destroy the clutches of eggs. After hatching, the canes are placed in the centres of the nests to mark the points to be surveyed. We have usually surveyed the colony towards the end of the season when most of the field work has been done.

The season of field work extends from mid-May when the first eggs are laid until mid-August when the last chicks fledge. By mid-September all the birds have left. About six weeks separate the breeding of the first and last pairs. The first pair to breed, pair number 106, usually hatched their first egg on 12 June. The last hatching date has been 22 July. Several pairs have been as late as 20 July. These very late pairs always consist of two young birds breeding for the first time. The general distribution of breeding dates is very skewed. Most pairs have bred together previously. They hatch their first egg at a mean of about 22 June. Very few of these old pairs hatch their eggs after the end of June. New pairs hatch their first egg around a mean date of about 1 July. They produce a long tail of late breeders. The distribution of breeding dates is thus truncated at about 11–12 June (no pair has been earlier than 11 June); the median date is 23 June and there is a long tail of late breeders extending to about 20 July. New pairs start hatching their eggs from 22 June onwards. Their eggs were laid from the last week of May onwards. Since birds can only be caught while incubating eggs, the sequence of dates when the nests were found must be used to determine the latest dates when the eggs will probably hatch and hence the latest dates when any unringed birds may be caught. Similarly, the hatching dates are used to determine the dates when the chicks should be caught for ringing. A chick can be ringed safely at about 14 days old, but we generally aimed to catch the chicks at 21 days old when they could easily be classified as either pale, intermediate or dark. They fledge at about 27–28 days which is therefore the latest they can be caught. At this age, they can easily be classified according to phenotype. The chicks do, however, differ considerably from adults in their phenotypic variations. Intermediate chicks appear to overlap phenotypically with both pale and dark chicks, whereas intermediate and pale adults are completely distinct. This suggests that our classifications of chicks and adults do not exactly correspond. I shall discuss this problem further when considering the genetics of the polymorphism (chapter 5). In the next section, I describe the characters we use to classify the phenotypes.

1.4 Classifying the phenotypes

We have recognized four phenotypes, both of adults and chicks: pale, intermediate, dark intermediate, and dark. The classification of intermediate and dark intermediate is arbitrary: variation is continuous from one phenotype to the other. Dark intermediates are not completely distinct from darks: some individuals may be placed arbitrarily in one class or the other. The pale adults, however, are completely distinct: although they do vary among themselves to some extent, they are quite different from the palest intermediates. An initial impression thus suggests that dark may be a genetically semi-dominant character, the heterozygous intermediates overlapping in phenotype with the homozygous darks.

The phenotypes of adults

I have already described the general field characters of the adult phases in the first section of this chapter: the pales, with their white underparts and neck; the intermediates with a pale neck; and the uniformly dark brown darks. As we have seen, the pale adults are always completely distinct from the other phenotypes. But they vary in the colour and extent of their pale plumage. Some pales have a completely white belly, breast and neck. Only the wings, tail, back and crown of the head are dark brown. Such birds are strikingly handsome and conspicuous. In most pales, a band of somewhat darker feathers across the breast separates the white of the neck from the belly. These birds also tend to be less dazzlingly white on their underparts. The white feathers may be flecked with light brown at the tips, though the feathers are almost always pure white in the middle of the belly. A few birds present a largely flecked appearance on their pale parts. At a distance they may appear pale fawn rather than white. These pales have been called 'dusky pales'. But their pale chicks have always had the normal characteristics of pale chicks. Dusky pales appear to represent just the extreme end of the range of variation in the expression of the pale phenotype. They seem to retain some of the flecking which is always present in chicks but which is usually lost by the time the birds have reached breeding age.

Intermediate adults vary widely, like the pales. But since they are very close to the darks in the average expression of their phenotype, their range of phenotypes overlaps that of the dark phenotype. Their paler neck is the normal field character which identifies them. It may be as pale as the neck of one of the darker pale phenotypes. Or it may be dark except for the presence of some straw-coloured or golden-coloured feathers on either side of the neck. This gives the head a 'capped' appearance from the somewhat lighter appearance of the neck compared to the nape and crown. But when the bird extends its neck, the paler feathers are spread out and less distinctly seen. We would call these birds dark intermediates. The paler, more obviously intermediate birds, may often have a somewhat less uniformly dark belly. In intermediates, the dark belly feathers have white bases. In some of the paler intermediates, the belly feathers may be white from their base to about half their length. The white bases of the feathers tend to lighten the general appearance of the belly. We have used the white bases of the belly feathers as the diagnostic character of intermediate phenotypes. It can only be distinguished on a bird in the hand. Even so, the distinction is not always clear: some birds have dark belly feathers that lighten in colour towards the base without necessarily having a clear white base. These birds may have only a few paler feathers on the sides of the neck. We would call them dark intermediates, but their characteristics merge into those of the dark phenotypes. Indeed, a bird classified as dark intermediate in one year may appear as dark in a later year, or a dark in one year may later appear as dark intermediate. Clearly the annual moult gives rise to some slight phenotypic variation in individuals, though a bird that is clearly intermediate with an obviously paler neck will always retain this character. As we shall see, the pales always produce pale chicks from pale × pale matings. Similarly dark × dark matings almost always produce dark chicks. This of course immediately suggests that pales and darks are genetically homozygous for the alleles that determine these phenotypes. The intermediates would consequently be heterozygous. Unfortunately, intermediates form the greater proportion of individuals in the population – about 60 per cent. This is genetically impossible (see section 5.2, chapter 5). It implies some of the birds we classify as dark intermediate are probably homozygous like the darks. We should not expect that phenotypes would correspond exactly to the genotypes, and they do not appear to do so.

The phenotypes of chicks and fledglings

All chicks show some flecking at the tips of their feathers. The white belly feathers of the pale chicks are flecked with light brown except in the middle of the belly where a patch of pure white feathers is always found. This is the diagnostic character of a pale chick. The dark brown feathers of the wings and back are also flecked in the chicks. In the pale chicks, these feathers have an extensive light brown bar across the tip. From a distance, the chicks appear mottled light and dark brown. Instead of the white neck of adult pales, the chicks have a neck of buff-coloured feathers. As they get older, they generally become paler, for the bases of the neck feathers are generally white. The fledgling is a very handsome bird, (see Figure 1.4), strongly marked with orange-brown bars on both the dark wings and back and on the white neck and belly. This barring seems to persist in one-year-old birds and also in some two-year-old birds. Immature pale birds with strongly barred underparts can often be seen among groups of non-breeders; some pales breeding for the first time may still retain a slight barring on their belly feathers.

Intermediate chicks are classified by the same criterion as the intermediate adults: they have dark belly feathers with white bases. The white base may extend along as much as half the length of the feather. The dark end of the feather may be divided into two dark bands, separated by a pale band. The feathers of the wings and back usually have light brown tips, like those of the pales, but the width of the light brown band is smaller. Intermediate chicks thus appear less strongly barred than pale chicks. They are much more 'intermediate' between pale and dark than the intermediate adults. No intermediate adult approaches the adult pale phenotype in its degree of expression. Some intermediate chicks do closely resemble the pales: they are separated only by the criterion that pale chicks have a patch of pure white belly feathers, whereas intermediates always have dark tips to these feathers. Even this criterion is not completely sharp: all gradations can exist

Figure 1.4 A pale chick; its belly feathers shown in close up.

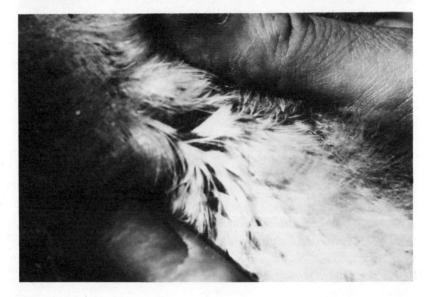

from a white patch of feathers, to just a few pure white feathers, to belly feathers all of which have at least some indication of a darker tip, and so on. Figure 1.5 shows the typical intermediate chick, with barred back, resting on the ground; its belly feathers in close-up. Its barred plumage helps to conceal it against the background of the heather.

Similarly the intermediate chicks can merge phenotypically with the darks. Dark chicks have dark brown belly feathers with no white base (Figure 1.6). This usually produces a clear distinction. But some intermediates have only a narrow white base to the feather. At some stage, this could presumably become so narrow that no obvious distinction could be drawn. Dark chicks are not completely dark on the wings and back: usually they show slight barring, though some may be almost unbarred. Intermediate chicks are thus much closer to the middle of the spectrum between darks and pales than intermediate adults. The phenotypic range of the intermediate chicks is wide enough to merge at both extremes into the ranges of both dark and pale phenotypes. But adult intermediates never approach the pale phenotypes in expression. On the basis of the criteria I have defined, the amount of overlap and misclassification is very small. But this applies only to the phenotypes defined by these criteria. It does not follow that the genotypes are not being misclassified. For example, if intermediate chicks were usually heterozygous, we should expect that, as they became much more similar to the darks in adulthood, the genotypes would tend to be misclassified: intermediates would sometimes be homozygous; darks would sometimes be heterozygous; the differences between the phenotypes would no longer separate the genotypes. As we shall see in chapter 5, genetic evidence can be used to determine the probable extent of any such misclassification.

1.5 Ecology and genetics of Arctic Skuas

Melanic forms are found in many species of birds. The Arctic Skua is one of the most striking examples, since the non-melanic pale phenotype is so clearly different from the melanic intermediate and dark. The blue melanic form of the Lesser Snow Goose is also very different from the non-melanic phenotype. In other cases of polymorphic melanism, the phenotypic differences are less extreme. Fulmars have a slightly darker 'blue form' which is very rare in the southern parts of its range but becomes somewhat commoner towards the north. Two Mediterranean falcons, *F. eleonorae* and *F. concolor* produce melanic phenotypes. Many pigeons are melanic in urban populations.

The melanism of the feral pigeon can be explained in terms of an association between the breeding cycle and pituitary hormone levels.

Murton & Westwood (1977) suggested that melanics have an altered photoresponse: this is indicated by their lengthened breeding season; they may continue breeding during the winter. The melanic males develop larger gonads and produce more spermatocytes than the non-melanic wild-type

Figure 1.5 An intermediate chick; its belly feathers shown in close up.

pigeons. In a considerable proportion of melanics, the gonads do not regress at the end of the normal breeding season. They remain continuously in breeding condition (Murton, Westwood & Thearle, 1973). The melanics' breeding season thus extends into the period of moult.

Figure 1.6 A dark chick; its belly feathers shown in close up.

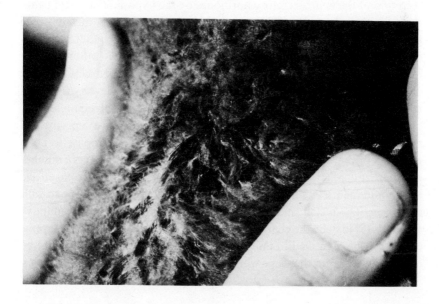

Since breeding and moulting use a considerable proportion of available energy, this would normally be highly deleterious, in spite of the reproductive advantage the melanics may obtain. But melanics are common only in towns. Murton *et al.* (1973) studied a polymorphic population containing a high proportion of melanics that lived in grain silos in Salford, Manchester. Of course, unlimited high energy food was available to the birds of this population. No doubt they could afford the energy both to moult and to reproduce at the same time. As expected, the melanics were reproductively more successful: they produced significantly more offspring than the non-melanics. They also mated disassortatively: wild-types preferred to mate with melanics; blue melanics preferred to mate with dark-blue melanics (see section C.4.2, appendix C). Preferential mating strongly favoured melanics. The melanics thus gained a sexual advantage. It may be presumed that this is offset by the disadvantage of continuous breeding unless unlimited food is available to supply the additional energy.

But what is the connection with melanism? The extended or continuous breeding season may obviously be an adaptation to unlimited food supply. Why is it associated with melanism? Murton & Westwood (1977) speculate that the melanism is the incidental effect on plumage of a gene that primarily alters the bird's photoperiodism and hence the timing and duration of the breeding season. Such incidental effects are said to be the 'pleiotropic' effects. A gene which alters the periodicity of a circadian oscillator will alter the photoresponse to daylength, hence the timing of secretion of the neurosecretory hormones by the hypothalamus and hence the timing of secretion of the anterior pituitary hormones including the gonadotrophins (luteinizing hormone, LH, and follicle-stimulating hormone, FSH) and the melanophore-stimulating hormone (MSH). Earlier induction of LH at short daylengths will produce an earlier recrudescence of the gonads and their later regression. Higher average levels of LH will produce greater development of the gonads during the breeding season. Similar effects on MSH will also produce the more extensive deposition of melanin in these individuals. Melanism will thus be a pleiotropic effect of a gene that primarily alters the photoresponse. A gene often affects a number of different characteristics of an organism. Some of its pleiotropic effects may not in themselves have any adaptive significance. This may be so with melanism. The melanism itself may be neutral or of only slight adaptive value. But it may be the characteristic by which a gene with other adaptive effects can be identified in individuals. Of course the melanism might be closely linked to another gene which determines the adaptation. Pleiotropy of gene effects cannot easily be distinguished from close linkage. But close linkage would only produce such an apparent pleiotropic effect if, let us say, the gene for melanism were always in combination with the gene which altered the photoperiodism, so that the melanism were always associated with earlier breeding and an extended breeding season. This association would be very unlikely to persist in the long run. Eventually, by recombination, the genes would associate at random: the gene for melanism would occur as often in combination with the gene for normal breeding as with the gene for extended breeding. A particular combination of genes could be maintained only if both genes were strongly selected because of their combined effect. For example, the combination of melanism with extended breeding could only be maintained if the melanism itself were to provide a large part of the advantage of the extended breeding. If the melanism itself has only a slight selective effect, the combination could not be maintained: melanism would occur as often with normal breeding as with extended breeding. Very close associations of linked genes are indeed only found in cases where the genes interact strongly to produce selection in favour of specific combinations. A more or less random combination of genes is the general rule. Melanism is most likely to be a pleiotropic effect of the gene that alters the photoperiodism.

Darwin developed his theory of evolution to explain adaptation. Although much of the evolutionary change in proteins is probably non-adaptive (see Nei, *Molecular Population Genetics and Evolution, 1975*), the morphological and physiological changes by which the major groups and species of organisms have come into being can only be explained by adaptive evolution. These changes may well have carried with them pleiotropic effects which are not in themselves adaptive. If we can only recognize the pleiotropic effects that produce the different phenotypes, we shall not necessarily be able to determine what the adaptive character is or what ecological factor it is an adaptation to. The examples of melanism and mimicry in insects are always chosen to illustrate the meaning of adaptation precisely because the ecological factors that determine their selection have been isolated and manipulated experimentally. It is known how the genetic effects determine the fitness of the genotypes. In the melanism of the moth *Biston betularia*, the dark melanic phenotypes were found to occur at high frequency only in areas of atmospheric pollution. By itself, this observation gave no hint why melanism should be an adaptation to atmospheric pollution. Further observation showed that the moths rest on tree trunks which are black in polluted areas but encrusted with silvery lichens in unpolluted areas. Birds have been shown to take melanics preferentially from lichen-covered tree trunks in country areas and non-melanics from bare black tree trunks in industrial towns (Kettlewell, 1973). Experiments by Clarke & Sheppard (1966) and Bishop (1972) compared the relative predation rates on different trees in different areas. Melanics had the higher survival rates on dark tree trunks; non-melanics on the white tree trunks of silver birches. These differences in survival rate were used to calculate the relative fitness of the melanics and non-melanics in the two different environmental backgrounds. Melanism was thus shown to be an adaptation to two environmental factors: the presence of birds as predators and the presence of black tree trunks caused by atmospheric pollution that had killed the lichens. When these factors were present, the melanics had the higher fitness because the predators were less likely to see them than the non-melanics.

In the study of the ecological genetics of melanism in moths, the fitnesses of different genotypes were measured and shown to be determined by certain factors that could be identified in the environment. Melanism in birds may also be an adaptation in itself. Arnason (1978) found that pale birds were more successful in their attacks on prey than dark birds. He suggested that the prey might be looking out for attacks from the much commoner dark birds, thus giving advantage to the rarer pales – a form of apostatic selection. Predators of Arctic Skuas, like Bonxies or Arctic Foxes, may possibly be able to find some phenotypes of chicks more easily than others. These are hypothetical possibilities which could only produce very weak selection. Arctic Skuas have no difficulty obtaining food and spend very little time hunting. By my calculations, Arnason's results are not statistically significant (see chapter 3). Except in areas where Bonxies and Arctic Foxes are very common, the Arctic Skuas themselves seem to have no predators. It is much more likely that the melanism of the Arctic Skua is but the pleiotropic effect of a gene that primarily determines physiological or behavioural adaptations.

The fitness of different phenotypes or genotypes in a population can be measured without knowing what factors in the environment produce the variation in fitness. A population genetic analysis on its own cannot show what factors in the environment melanism is adapted to: it can show how melanism evolved and why it is maintained in a population as a stable polymorphism.

1.6 Relationships with other species of Skuas

In the Northern Hemisphere, the Arctic Skua co-exists with three other species of skuas: the Great Skua or Bonxie, *Catharacta skua*; the

Table 1.1. *Comparative measurements of Skuas*

	Catharacta species	*Stercorarius pomarinus*	*Stercorarius parasiticus*	*Stercorarius longicaudus*
Wing (mm)	380–405	340–373	305–328	290–320
Central tail feathers (mm)	140–155	165–225	170–215	275–355
Outer tail feathers (compared to central)	10–25 mm shorter	50–105 mm shorter	55–105 mm shorter	160–250 mm shorter
Tarsus (mm)	65–72	48–56	42–45	35–44
Bill (mm)	47–55	35–40	27–32	25–30

Pomarine Skua, *Stercorarius pomarinus*; and the Long-tailed Skua, *Stercorarius longicaudus*. Two species of skua inhabit the Southern Hemisphere. One of these, the Southern Skua, is very similar to the Bonxie. It is usually considered to belong to the same species and given the same specific name, *C. skua*. The other species, McCormick's Skua *Catharacta maccormicki*, is much paler though otherwise similar to the Bonxie and Southern Skua in size and morphology.

These skuas are all members of the family *Stercorariidae*. They are all gull-like birds that feed as predators and pirates. They differ from gulls in having a sternum with only one pair of incisions, not two; they have a hooked bill like raptors and a cere; their tarsi and feet are covered with rather more prominent scutes than gulls. Like raptors and unlike gulls, they have brown plumage. Three species are polymorphic, all with very similar melanic and non-melanic phenotypes. These polymorphic species, the Arctic, Pomarine and Long-tailed Skuas, also possess the very elongated central tail feathers characteristic of the genus *Stercorarius*. The skuas obviously divide into two groups: the *Stercorarius* group and the *Catharacta* group. The pale non-melanic phenotype with white underparts and neck is confined to the *Stercorarius* group; it must have arisen by mutation after *Stercorarius* became separated from *Catharacta*.

The four species of skuas in the Northern Hemisphere show much greater divergence than the two species in the Southern Hemisphere. The *Stercorarius* species in the Northern Hemisphere show more divergence than both Northern and Southern species of *Catharacta*. The *Stercorarius* species diverged in size: the Pomarine Skua is the largest and heaviest; the Long-tailed Skua is the smallest and lightest. *Catharacta* species are all similar in size. Table 1.1 gives the comparative measurements. On morphological grounds, the evolutionary relationships are easily deduced: *Stercorarius* must first have separated from *Catharacta*; the three species of

Stercorarius then diverged, the small Long-tailed Skua evolving farthest and diverging most from the ancestral *Stercorarius* species; finally *Catharacta maccormicki* diverged from *C. skua*. Figure 1.7 shows the evolutionary tree of these relationships.

The skuas must have had their origin in the Northern Hemisphere, the Southern skuas having diverged at a comparatively late date within the *Catharacta* group. In the Southern Hemisphere, *C. maccormicki* differs in its geographic range from *C. skua*. McCormick's Skuas are confined to the Antarctic continent and South Shetlands. They breed close to rookeries of the Adélie Penguin *Pygoscelis adeliae*, feeding on penguin chicks and eggs and scavenging within the rookeries (Trillmich, 1978). They have been seen further south than any other species of bird. This extreme range and specialized feeding has presumably determined the differentiation of McCormick's Skua from the Southern Skua and Bonxie. The Southern Skua is distributed over similar Southern latitudes to the Northern latitudes of the Bonxie's distribution.

Within the *Stercorarius* group, the Long-tailed Skua is differentiated from the Arctic and Pomarine Skua by feeding behaviour and the frequency of its dark phenotype as well as by morphology. It lives less by piracy than the two other species. In the breeding season it feeds on lemmings, mice, birds' eggs, insects, crustaceans, earthworms and berries. The Long-tailed Skua differs from the two other species in the frequency of its dark phenotype. Very few dark Long-tailed Skuas have ever been recorded, though it breeds in Norway where 50 per cent of Arctic Skuas are dark. Like the Long-tailed Skua, the Pomarine Skua has a more northerly distribution than the Arctic Skua. Dark Pomarine Skuas occur at a

Figure 1.7 Evolutionary tree of the Stercorariidae.

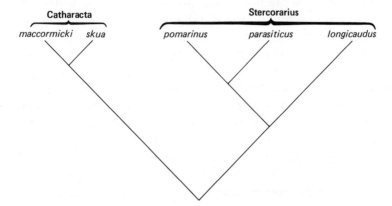

frequency of about 14 per cent. They are rarer than dark Arctic Skuas at the same latitudes, for example in Iceland and Greenland, but much commoner than dark Long-tailed Skuas. The data of phenotypic frequency, feeding behaviour and morphology all support the evolutionary tree shown in figure 1.7.

2

Numbers and distribution

2.1 Surveys of breeding colonies

Like many birds, Arctic Skuas can be observed and counted fairly easily on their breeding grounds. Breeding birds have been counted at many different colonies in the last 50 years. After the breeding season is over, they can be counted at points along their routes of migration. Counts of numbers of migrating birds give rough estimates of the relative proportions migrating along different routes, but no estimates of population sizes. Even in a breeding colony, numbers are difficult to estimate accurately. There are several sources of error. Estimates are usually too low. The exact number of breeding birds in a population can be determined only by surveying and marking all the nests as they are found. The survey must be repeated every two or three days throughout the period when the birds are incubating their eggs.

On Fair Isle, the first eggs are usually laid on 16 May. The last eggs may be laid on about 25 June, 40 days later. If a survey were carried out at the end of May, little more than half the pairs would be found. If a survey were carried out on about 20 July, when the last eggs were hatching, about half the pairs would already have fledged their chicks. They would no longer be defending their territories or feigning injury at the approach of an intruder. Many of the pairs would then be very difficult to ascertain.

Observers have usually counted Arctic Skuas by walking through the breeding colonies counting birds as they go along. They have counted either the individual birds or the pairs. Counting pairs is the more reliable method, particularly in dense colonies where many birds are whirling about the sky in dog-fights. Has a particular bird been counted already, or not? It may often be impossible to say. If a pair has eggs or young chicks, they usually feign injury when intruders on their territory approach their nest or chicks. Different pairs can usually be distinguished from each other as an observer passes from one pair's territory to another's: the birds of one pair will be seen to feign injury, then the birds of the other pair. But many pairs are inevitably missed in a single survey. Some, as I mentioned in chapter 1, are exasperatingly undemonstrative: they retreat to leave their territory undefended. In very dense colonies, such as the colony on the island of Foula in Shetland, different pairs may be difficult to distinguish in the general melée. When nests are close together – in places on Foula no more than 10 m apart – birds incessantly invade each other's airspace over their territories. The aerial dog-fighting becomes almost continuous. An observer walking through the colony puts up birds from their nests, creating further confusion. It becomes impossible to say how many different pairs have been seen.

A survey of birds thinly scattered in loose colonies over wide areas of moorland should reveal most of the pairs with eggs or chicks. Even so, only one bird of a pair may be present on the territory; the other bird may be away hunting. Single birds displaying distraction behaviour or defending a territory should obviously be counted as a pair. If individual birds are counted, the absent birds will obviously be missed.

Surveys should be carried out in the middle of the breeding season. In Shetland, almost all pairs will have laid their eggs by 20 June. The first fledglings will not be flying before 12 July. Between these dates all territories should be occupied. Every pair then has a chance of being found and recorded in the sample. Bengtson & Owen (1973) surveyed numbers of Arctic Skuas in June and July 1971 while travelling over most of the coastal areas of Iceland. They counted every bird they saw at close range and attempted to score their phenotypes as melanic or pale. They identified the phenotypes of 557 birds. In their paper they state that a total of about 8000 Arctic Skuas breed in Iceland. They do not explain how they obtained this estimate.

In my own experience, counting birds during a single survey produces a very low estimate of the true population size. Pairs are so easy to miss in densely colonized areas. Only by marking each nest and repeatedly suveying the colony throughout the breeding season can all the nests be found. On Fair Isle, starting on 1 June, I had usually found and marked about 80 per cent of the nests 10 days later. I found the remaining nests more and more slowly during the next 20 days, in the course of the business of nest checking. Once found, each nest is checked every one or two days to determine the clutch size and the hatching dates of the eggs. Later the chicks are ringed and checked for date of fledging. A complete check of 120 nests will take at least half a day to carry out. In the course of continually walking through the colony on nest checks, new pairs come to light as birds are seen to be present or to attack or to feign injury where no nest has been marked. To find new pairs, the marking of each nest is essential: a new pair showing aggressive or distraction behaviour can then be distinguished from known pairs displaying at the marked nests. New pairs are also observed from hides if a suspected nest cannot be found. Thus, gradually, in the course of the breeding season, new pairs and nests are located and marked. The last few nests may be found one by one in the course of many days spent nest checking. To get an accurate estimate of the population size, the whole breeding season must therefore be covered. Only one or two nests can have been missed in a season's field work on Fair Isle.

Foula is the only other Arctic Skua colony which has been intensively surveyed throughout the whole of the breeding season. John Davis spent two weeks on Foula in 1974 mapping nests and ringing adults. In the following year he and I both worked for three weeks on Foula assisted by Shoshana Ashkenazi. Shoshana remained on Foula for the whole of the breeding season after John Davis and I had returned to Fair Isle. A total of 278 pairs were eventually found and their nests marked. This excludes isolated pairs nesting along the east coast of the island. At least 300 pairs of Arctic Skuas must have bred on Foula in 1975. Figure 2.1 shows a map of the island, and the extent of the Arctic Skua colony. Previously, in 1974, John Davis had found about 240 pairs. These estimates are greatly in excess of the previous estimates of population size. In the years 1960–69 the Brathay Exploration Group had taken parties of children to Foula and surveyed the numbers of Arctic Skuas. Their estimates range from 100 to 160 pairs. In 1972, the number of pairs was estimated to be 150. In 1973 Furness (1977) estimated a population size of 130 pairs. It is fair to assume that all these estimates were far too low. An increase in actual numbers from 130 pairs in 1973 to 300 pairs in 1975 would represent an exponential rate of increase of 52 per cent in each of two years after a long period of stability. The population size in the years up to 1973 had obviously been under-estimated by a large factor.

The surveys that have been carried out on Foula for many years may be of little use even for comparative purposes. We have already seen that a survey must be carried out at the end of June or beginning of July if all pairs are to have a chance of being found. Even if each survey had been made at the same time each year, some surveys may have been carried out over longer periods than others or carried out more intensively. Different proportions of the total number of pairs would then have been sampled. To determine population trends over a number of years, exactly the same sampling method must be used between the same dates in each year. On Foula, pairs have sometimes been counted, sometimes mapped. Mapping, as we should expect, has always produced larger estimates of population size. I have no information on the exact dates of the surveys, the numbers of individuals co-operating in the survey, or the details of how the work was done. The variation in the estimates between 1948 and 1973 could obviously have arisen by variation in the sampling method. I suspect that the true population size was roughly twice the estimates obtained in these years.

All the main Arctic Skua colonies in Orkney and Shetland have been

Figure 2.1 Map of Foula. The stippled areas show the main nesting grounds of the Arctic Skua. The solid black diamonds indicate isolated pairs. Enclosed crofting areas are shown with name of croft.

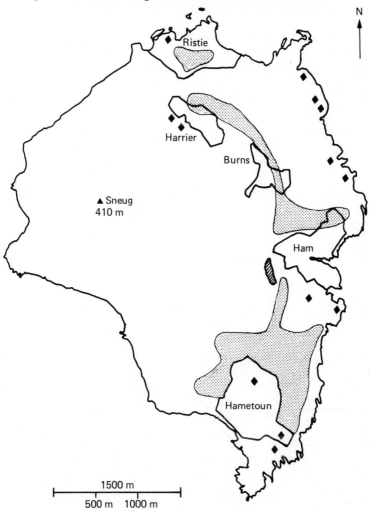

surveyed at some time. Only those populations on Fair Isle, Noss and Foula have been surveyed intensively for many years. The map in figure 2.2 shows where the main colonies of Arctic Skuas are found in Britain. The true size of the Fair Isle colony is known for each year from 1948 to 1962 and from 1973 to 1979. The 300 pairs we found on Foula must be close to the true population size. The population on Noss is also known exactly. Wardens of the Bird Sanctuary map the nests in the course of each breeding season. About 40 pairs usually breed each year. From 1969 to 1974, the population has varied between 40 and 44 pairs (Harris, 1976). In this small population, close to the Warden's cottage, pairs are unlikely to have been missed. Elsewhere in Orkney and Shetland, most populations have been surveyed at some time. These casual surveys, some carried out a long time

Figure 2.2 Map of Shetland, Orkney and Scotland showing the positions of the main Arctic Skua colonies. Solid black circles indicate order of magnitude of numbers of pairs.

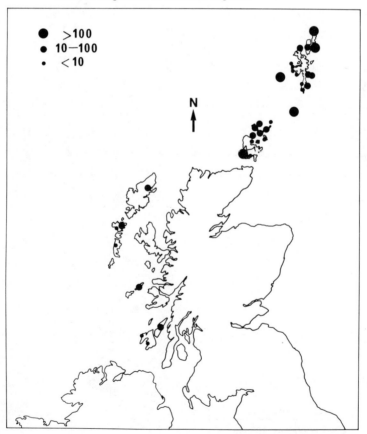

ago, must always have under-estimated the size of Arctic Skua populations. In 1969 and 1970 the 'Seabird Group' attempted to survey all Britain's seabirds (Cramp, Bourne & Saunders, 1974). Many different observers surveyed colonies at various times. The results can hardly be comparable because of differences between both the observers and the times when the colonies were surveyed. Putting all the data together, Cramp, Bourne & Saunders estimate that a total number of 1090 pairs of Arctic Skuas breed in the British Isles. Since they used an estimate of 120 pairs breeding on Foula, their total estimate is likely to be only one-third of the true number.

Arctic Skuas have been counted in Iceland, Finland and Faeroe. Bengtson & Owen's estimate of 8000 birds for the population in Iceland is perhaps more of a guess than an estimate. It was clearly not a total number of birds actually counted (Bengtson & Owen, 1973). Hilden (1971) surveyed the Arctic Skua populations of Finland in great detail, counting a total of only 225 pairs. He suggested that the error in this figure was unlikely to exceed 10 per cent but did not explain how he estimated this error. Elsewhere in the Northern Hemisphere, Arctic Skua populations seem to have been surveyed in detail only in the Faeroes. The total numbers of birds seen were counted in several different parts of Faeroe. Southern (1943) assembled data from all the records of Arctic Skuas he could find. A total of 394 pairs were counted in different parts of Faeroe. Large colonies were said to exist in other parts. On this basis, the Faeroe population must be a large one. Arctic Skuas breed throughout the coastal regions of the Northern Hemisphere between the latitudes of 60°N and 80°N. The map in figure 2.3 shows the approximate limits of their breeding range. They are certainly common in Alaska, for large numbers have been seen on migration (Dean, Valkenburg & Magoun, 1976). They are reported as occurring widely in Greenland, Spitzbergen and the USSR. The reports give no idea of any population sizes. The world population of Arctic Skuas cannot even be guessed at to any order of magnitude.

2.2 Changes in population numbers and status

Seabirds in the British Isles appear to be either increasing or fairly stable in numbers. The auks – Razorbills (*Alca torda*), Guillemots (*Uria aalge*) and Puffins (*Fratercula arctica*) – are probably fairly stable in numbers, but they are all so difficult to count that figures of population sizes are meaningless. Large new colonies of Guillemots have been formed in some areas, and their numbers may therefore be increasing (Harris, 1976). Fulmars, Great Black-backed Gulls and Gannets, which are easier to count, are still increasing in numbers (Harris, 1976). Gannets are still colonizing new areas. The new colony on Fair Isle is expanding rapidly. But the most dramatic increases are those that have taken place in the numbers of the Great Skua or Bonxie. From having been reduced to a few pairs on

Figure 2.3 Map of the Northern Hemisphere showing the rough limits of the Arctic Skua's range.

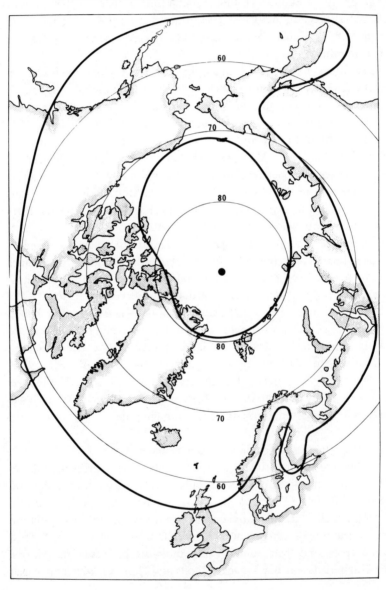

Table 2.1. *Estimates of pairs of Arctic Skuas on Foula, Shetland*

Year	Source/observer	Method	Pairs
1890	Barrington (1890)	counting pairs	60
1948	Pennie (1948)	counting pairs	100
1960	Brathay (unpubl.)	mapping pairs	131
1962	Brathay (unpubl.)	counting pairs	120
1964	Brathay (unpubl.)	counting pairs	120
1965	Fabritius	mapping pairs	140
1968	Brathay (unpubl.)	mapping pairs	160
1972	Brathay (unpubl.)	mapping pairs	150
1973	Furness (1977)	mapping pairs	130
1974	Furness (1977)	mapping pairs	180
1975	Davis and O'Donald	marking nests	278

Brathay (unpubl.) refers to observations made by the Brathay Exploration Group. In his paper, Furness refers to J.W.F. Davis 1975 as marking 253 nests. In fact John Davis and I marked 278 nests in our study of the mating frequencies and genetics of the Arctic Skuas on Foula. We did not mark the nests of about 20 isolated pairs breeding along the East coast of Foula. The total number of pairs is probably about 300. I have no idea where Furness obtained his figure of 253 marked nests in 1975.

Foula and Hermaness in Shetland at the end of the last century, Bonxies had increased by 1974 to about 2500 pairs on Foula and 800 pairs on Hermaness. They are found in smaller numbers on many other islands and parts of Shetland mainland.

Changes in Arctic Skua populations have been less dramatic. As we have seen, counts of Arctic Skuas, like counts of other seabirds, are unreliable unless each nest is found and marked. This method has been used only in surveys of colonies on Fair Isle, Foula, Noss and Hermaness. Only on Fair Isle has it been used exhaustively throughout the breeding season to find all the nests. Furness (1977) compiled data of numbers of Arctic Skuas on Foula, Noss and Hermaness. These data and the data of the population on Fair Isle are shown in tables 2.1, 2.2, 2.3 and 2.4. The tables indicate the method of counting together with references to the sources of the data.

I suggested in the previous section that the earlier counts on Foula must have greatly under-estimated the number of breeding pairs, for an increase from 130 pairs in 1973 to 300 pairs in 1975 is incredible. The estimates of pairs on Foula do show a general upward trend. Brathay's mapping gave

Table 2.2. *Estimates of pairs of Arctic Skuas on Noss, Shetland*

Year	Source/observer	Method	Pairs
1922	Baxter and Rintoul (1953)	counting pairs	45
1929	Perry (1948)	counting pairs	60
1934	Perry (1948)	counting pairs	50
1939	Perry (1948)	counting pairs	25
1946	Perry (1948)	marking nests	31
1947	Baxter and Rintoul (1953)	counting pairs	37
1955	Kinnear (N.C. Report)	counting pairs	25
1958	Kinnear (N.C. Report)	counting pairs	25
1969	Kinnear (N.C. Report)	counting pairs	40
1970	Kinnear (N.C. Report)	counting pairs	40
1973	Kinnear (N.C. Report)	mapping nests	39
1974	Kinnear (N.C. Report)	marking nests	44

Kinnear (N.C. Report) refers to Peter Kinnear's observations which he describes in a Report to the Nature Conservancy Council. N.J. Gordon also carried out surveys of Arctic Skuas for the Nature Conservancy in 1957 and 1964, counting 13 and 17 pairs.

estimates of about 130 pairs in 1960 and 160 pairs in 1968, rising to Furness' estimate of 180 pairs in 1974 and our own estimate of 300 pairs in 1975 obtained by marking all nests found during the breeding season.

The small colony on Noss, situated close to the Warden's cottage, is easily counted. Kinnear's maps of the nests give reliable estimates of the numbers of pairs. The colony has shown no increase in recent years. Between 1922 and 1934 it seems to have been larger than it is now, for the less reliable method of simply counting pairs gave larger estimates of population size in those earlier years.

The population on the Nature Reserve of Hermaness on the island of Unst has certainly declined. The Bonxies, which increased from three pairs in 1900 to 800 pairs in 1974, have simply pushed the Arctic Skuas off their former territories. The Arctic Skuas have moved away from the Nature Reserve and now nest as isolated pairs or in loose colonies along the West coast of Unst south of Hermaness. Bundy in a report to the Royal Society for the Protection of Birds counted 121 pairs of Arctic Skuas elsewhere on Unst – a total of 193 pairs including those still nesting on Hermaness near the Nature Reserve. This estimate suggests that the Arctic Skuas have moved their territories to get away from the Bonxies, but have not obviously declined in total numbers.

Table 2.3. Estimates of pairs of Arctic Skuas on Hermaness, Unst, Shetland

Year	Source/observer	Method	Pairs
1922	Pitt (1922)	mapping pairs	200
1937	Baxter & Rintoul (1953)	counting pairs	100
1950	Gordon (N.C. Report)	marking nests	75
1958	Gordon (N.C. Report)	counting pairs	70
1959	O'Donald (1960*b*)	counting pairs	142
1965	Dott (1967)	mapping pairs	60
1969	Bourne & Dixon (1974)	counting pairs	80
1974	Bundy (RSPB Report)	mapping pairs	72

Gordon (N.C. Report) refers to a Nature Conservancy Report on the Hermaness National Nature Reserve. Bundy (RSPB Report) refers to a report to the Royal Society for the Protection of Birds.

Table 2.4. *Numbers of pairs of Arctic Skuas on Fair Isle, Shetland*

Year	Pairs	Year	Pairs
1948	15	1973	106
1949	20	1974	116
1950	22	1975	137
1951	26	1976	136
1952	32	1977	139
1953	34	1978	114
1954	34	1979	114
1955	44		
1956	52		
1957	56		
1958	61		
1959	65		
1960	65		
1961	67		
1962	71		

On Fair Isle the increase in numbers of Arctic Skuas has been as dramatic as the general increase in Bonxies. The Bonxies have been increasing at an exponential rate of about 11 per cent per year. The data in table 2.4 show that Arctic Skuas also increased exponentially at 12 per cent per year.

Between 1948 and 1962, the numbers of pairs are given by the equation

$$N_t = 16.62 \exp (0.1092t)$$

where 1948 is the year at $t=1$ and 1962 is the year at $t=15$. Thus the percentage increase in one year is $100(e^r-1)=12$ per cent. For goodness of fit to the data, we obtain

$$\chi^2_{13} = 8.5911$$

corresponding to

$$P = 0.803$$

a very good fit indeed. Even so, the calculated numbers of pairs are greater than observed numbers in the years 1961 and 1962. For these years the equation predicts populations of 76.6 and 85.5 pairs compared with the 67 and 71 pairs that actually bred.

The equation predicts that by 1973 the population should have increased to 255 pairs; but it had only reached 106 pairs. The rate of increase thus declined sharply after about 1960. This is shown in figure 2.4. Taking the years up to 1960, the exponential rate of increase was 0.139 per year. From 1960 to 1973, it was 0.0383. It increased briefly again in the years from 1973 to 1975. The population then stabilized. After 1977, it started to decline. The decline is easy to explain. A number of birds were apparently shot in 1977 and 1978. Shooting may also have been the cause of the reduction in the rate of increase after about 1960. The Arctic Skuas were not intensively studied between 1962 and 1973. It would have been easier to shoot them

Figure 2.4 Rates of increase of pairs of Arctic Skuas on Fair Isle. The curves represent rates of exponential growth in three different periods as described in the text.

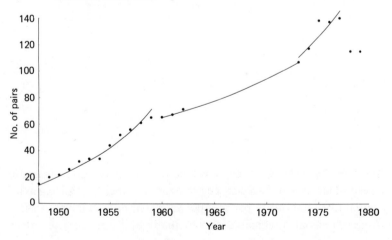

when observers were no longer working continually in the Arctic Skua colony. From 1973 to 1976 John Davis and I spent the whole of the breeding season working mainly on Fair Isle. After John Davis had left the project in 1976 I received some assistance from the wardens of the Fair Isle Bird Observatory, but the colony was no longer under the same intensive surveillance as in previous years. Until 1977 only one or two birds were shot each year. After 1977, twelve are known to have been killed. The human predation easily explains the decline in numbers on Fair Isle after 1976. In section 6.1.1 (chapter 6), I discuss the long-term effects of this predation. Bonxies have certainly driven Arctic Skuas away from Hermaness. The Arctic Skuas have now colonized other parts of Unst. They have not declined in total numbers. On Foula, in spite of the huge population of Bonxies – well in excess of 3000 pairs now terrorize large parts of the island – the Arctic Skuas may even be increasing. How the Bonxies and Arctic Skuas compete with each other is the subject of the next section of this chapter.

2.3 Regulation of numbers by predation, competition and food

Bonxies and Arctic Skuas obviously compete for territories. The Bonxies always seem to win. They are fearsome predators. They can take adult Auks, Herring Gulls and Kittiwakes in flight. They also feed as pirates, forcing Gannets to disgorge their food. Although swift in pursuit of prey, they are much slower than the extremely fast and manoeuvrable Arctic Skuas. When a pair of Arctic Skuas are chasing a Bonxie off their territory, they look like a pair of Spitfires attacking a slow, heavy Heinkel bomber. Bonxies cannot match Arctic Skuas in flight. But they have one great advantage in the competition for territories. They return to their breeding grounds earlier in the year. Table 2.5 shows the dates when the first birds were seen on Fair Isle and the first eggs were laid. The Bonxies are back at least two weeks earlier than the Arctic Skuas and lay their eggs about six days earlier. A pair of Arctic Skuas may return to find that Bonxies have taken over their territory. They can never dislodge Bonxies. They will repeatedly attack the Bonxies, often striking resounding blows, but the Bonxies always stand their ground. The rapid increase of Bonxies on Hermaness thus pushed the Arctic Skuas off the ground they formerly held. Eventually, the spread of Bonxies must restrict the breeding grounds of Arctic Skuas and many other birds. On Foula, this may already have occurred to some extent. Elsewhere, the Arctic Skuas have simply moved into vacant areas. On Foula, vacant areas of moorland favoured by Arctic Skuas (section 1.1, chapter 1) are becoming smaller. The Arctic Skuas

Table 2.5. *Dates when the first Bonxies and Arctic Skuas return to Fair Isle and lay their eggs*

Year	Dates when the first birds were seen		Dates when the first eggs were found	
	Bonxies	Arctic Skuas	Bonxies	Arctic Skuas
1973	3 April	16 April	14 May	18 May
1974	27 March	17 April	13 May	16 May
1975	31 March	16 April	—	16 May
1976	1 April	17 April	9 May	17 May
1977	1 April	20 April	11 May	19 May

appear to have been forced south from the airstrip to nest on lowland pastures, not their normal habitat.

Furness (1977) assumes that the apparent increase of Arctic Skuas on Foula (table 2.1) is real. If so, the exponential rate of increase of 0.52 per annum between 1973 and 1975 would be much too high to be explained by actual rates of adult mortality and numbers of chicks fledged (section 6.2, chapter 6). Furness thus concludes that the extra birds were immigrants. He supports his case for immigration on the evidence of the handful of immigrants (see section 2.4 of this chapter) trapped as adults on Fair Isle that were hatched and reared as chicks elsewhere in Shetland. About 45 per cent of adult birds recruited into the breeding population of Fair Isle appear to be immigrants (see section 2.4 of this chapter), but the numbers recruited into the Fair Isle colony showed no increase during the corresponding period of such apparently rapid increase on Foula. An increase in migration rate, such as that Furness has postulated, should have been observed throughout Shetland. To determine the rate of recruitment into a colony, all adults must be ringed and identified each year as on Fair Isle, so that new recruits can be distinguished from birds who have bred in previous years. I assume (section 2.2) that Arctic Skuas have indeed increased on Foula, but not by as much as the estimates of population size given in table 2.1. The largest estimates were derived from the most recent and most intensive surveys.

Although Arctic Skuas were probably still increasing on Foula up to 1975, they were also suffering in competition with the Bonxies. The Bonxies not only evicted them from their former territories but also preyed upon them. Table 2.6, taken from Furness (1977) shows the numbers of Arctic Skuas killed by Bonxies. These numbers are the totals of all remains of

Table 2.6. *Predation of Arctic Skuas by Bonxies on Foula, Shetland*

Year	Pairs of Arctic Skuas counted	Numbers killed by Bonxies	
		Adults	Fledglings
1969	100	20	51
1970	120	3	14
1971	140	11	72
1972	150	19	56
1973	130	3	43
1974	180	21	35
1975	300	17	26

Arctic Skuas that Bonxies had presumably killed and eaten. They must be fairly reliable figures, for Bonxies are the only predators that feed on Arctic Skuas in Shetland.

Between 1969 and 1973 (the years before the more intensive surveys by John Davis in 1974 and John Davis and myself in 1975), Bonxies ate 4.4 per cent of the total number of adults counted and about 31 per cent of the fledglings. Furness estimates the number of fledglings by assuming that on Foula the average production of fledglings per pair was intermediate between the less dense, more productive colony on Fair Isle and the denser, less productive colony on Noss. The mortality of chicks predated on Foula is about as great as the entire first-year mortality of fledglings from other colonies. The extra mortality and lower reproductive rate on Foula should eventually produce an annual 3.2 per cent decrease in numbers unless balanced by immigration from elsewhere (see Appendix A on estimating the intrinsic rates of increase of avian populations). Since numbers are still apparently increasing, predation cannot yet have become a factor regulating population size. Furness concludes 'Arctic Skua colony sizes are intrinsically regulated through control of recruitment in relation to food availability'. He accepts Wynne-Edwards' theory that organisms evolve behavioural mechanisms for regulating their populations to numbers at which they do not exhaust their food sources (Wynne-Edwards, 1962). This theory has never found general acceptance. It assumes that group selection favours populations with the appropriate self-regulating mechanisms, while other populations eventually become extinct. This assumption is quite implausible: first because if group selection occurred at all, it could take place only under very restrictive conditions of population size and structure; and second, because no mechanisms are known by which individuals can ascertain the size of their population and adjust their behaviour to maintain an optimum size. An ecologist would of course require the resources of modern technology – sampling schemes, computers, bomb calorimeters – to determine approximately what the optimum size of a population might be. Wynne-Edwards implicitly assumes that all this can be done by instinct. It is another of those sociobiological theories which assume that selection and genetic variation can do anything but which never explain in detail how the things are actually done. Characteristically, Furness (1977), in the sentence I have quoted, uses the term 'intrinsically regulated', thus completely evading the problem of giving any details of the mechanism.

Does the predation of Arctic Skuas on Foula require further explanation? Is it so great that the population would decline unless extra immigrants were recruited? If so, why should the migrants find their way to Foula but not to other Arctic Skua colonies? Furness implies that migrants have been drawn to Foula because Kittiwakes and Arctic Terns, which Arctic Skuas rob for food, have increased on Foula and more food would be available there. Yet these birds have also increased on Noss, Hermaness and Fair Isle (Harris, 1976). If juvenile Arctic Skuas go the rounds of the islands and assess the available food sources, as Wynne-Edwards' theory would imply, they should be attracted equally to Hermaness and Noss, where Arctic Skuas have stayed fairly constant in numbers. Finding food never seems to present any problem to Arctic Skuas on their breeding grounds. They spend only a short time – a quarter or half an hour – away from their territories to collect and bring back food for their mate and chicks. They spend most of their time just sitting and most of the remainder in defence of their territories. Other birds with young, especially passerines, are continually foraging. The clutch size of Blue Tits and Great Tits is selected for the maximum number of chicks the parents can successfully feed (Lack, 1954). Not so with the Arctic Skua: if one parent is killed, the other continues to incubate the eggs and rears the chicks successfully. Food cannot yet be a limiting resource to the Arctic Skuas in Shetland. On the other hand, the Bonxies must eventually limit Arctic Skua numbers. If Bonxies continue to increase in numbers and spread across Shetland, their predation and competition must become too severe and force the Arctic Skuas into a decline. On the very rough estimate of mortality on Foula, this point should already have been reached. But increased mortality of chicks will have a delayed effect on the adult population. The Bonxies increased from about 1100 pairs in 1968 to about 2500 pairs in 1974. The predation shown in table 2.6 occurred during this period of great expansion in the Bonxie population. The severe predation of the fledgling Arctic Skuas will only start to have an effect on the adult population four or five years later when the fledglings have come back to breed for the first time. If the increased predation and competition began in about 1968, recruitment would only have started to decline in 1973. If competition for territories were severe in the dense Arctic Skua colony on Foula, young adults would have difficulty in securing a territory. They would increase the size of the pool of non-breeding birds. If recruitment is initially into this pool, the pool would decline first and then the breeding population some years later. If we assume that the decline in fledgling production began in 1968, we should not expect to see a decline in recruitment until about 1975. This might not appear in the breeding population for another two or three years depending on the size of the pool of non-breeders. Thus the effects of predation might not be seen until about 1978 or 1979. Further observations on the Arctic Skuas of Foula are needed to test this hypothesis. A detailed demographic study, like that on Fair Isle, would be necessary to determine the rate of predation and its effect on the rates of increase or decrease of the Arctic Skua population.

2.4 Migration

Arctic Skuas migrate across generations from one breeding colony to another: a bird reared as a fledgling in one colony may breed in another. They also migrate every year, flying south from their breeding grounds in the autumn and returning in the spring. The annual migration can have no genetic effect on the populations. But migration across a generation from one population to another causes gene flow: genes from one population are introduced into another. Genes which are advantageous in one region may thus be transferred into populations in other regions where they are disadvantageous. Their selective elimination will ultimately be balanced by the immigration. When the balance has been reached, a gene will be common where it is advantageous, less common in areas where it has lost its advantage and become selectively neutral, and rare where it is disadvantageous. At some point in a neutral zone between areas of advantage and disadvantage, dominant and recessive phenotypes should occur at 50 per cent frequencies. This pattern of gene frequency distribution is called a cline. As we shall see in the next section, the melanic and non-melanic phenotypes of Arctic Skuas are distributed as a cline with frequencies varying from 100 per cent pale in the most northerly populations to about four per cent pale in some of the more southerly populations.

The migration across generations is difficult to estimate because only on Fair Isle are the chicks normally ringed. The number of adults from other colonies must be greater than the odd ones and twos that have come from Foula, Hermaness, Bressay, Mousa and Noss (a total of six immigrants between 1973 and 1979). One possible method of estimating the proportion of immigrants is to assume that all the chicks on Fair Isle have been ringed since 1973 when the present project on the Fair Isle skuas began. Any unringed birds coming to breed in the colony after 1978 (when birds fledged in 1973 would be six years old) would therefore have come from elsewhere. This is a safe assumption since most birds breed for the first time at three, four, or five years old and a few at six or seven years old. In 1979, 60 new birds were nesting in the colony. Of these, 26 had been ringed as chicks on Fair Isle. The remainder, 56.7 per cent, were presumably immigrants which had been reared in other colonies. This must be the upper limit for the proportion of immigrants since it assumes that all birds reared on Fair Isle will have been ringed. But we know that we have missed some of the chicks. If 10 per cent of chicks had not been ringed, the immigrants would represent 51.9 per cent of the new birds: if 20 per cent had not been ringed, 45 per cent would be immigrants. This is probably the most realistic figure. So far, all the known immigrants have come from colonies in Shetland. Other colonies of Arctic Skuas are much further away. The nearest of these, in Orkney, would probably provide any remaining immigrants. The high rate of about 45 per cent migration per generation that occurs between colonies in Shetland must produce some gene flow along the rest of the cline: genes that flow between adjacent colonies in one generation diffuse further into more distant colonies in subsequent generations. Eventually the rates of diffusion will reach a steady state. A stable distribution will then give the probabilities that a gene has diffused from different points along the cline. The diffusion may be sufficient to maintain the cline at an equilibrium in balance with selection that favours one gene in one region and another gene elsewhere. Clines can also be maintained by a balance between different selective pressures that vary from one region to another. Diffusion would then reduce some of the variation in gene frequency.

The annual migration of Arctic Skuas is unusual. Most seabirds simply spread out from their breeding colonies. The obvious exception is the Arctic Tern migrating far into the Southern Hemisphere. The Arctic Skua migrates as far as Africa: birds ringed on Fair Isle have been recovered on the Spanish coast, along the North African coast of Morocco and Algeria, and on the coast of Ghana. The map in figure 2.5 shows the points of recovery of Fair Isle birds. In late August and early September Arctic Skuas pass in considerable numbers down the east coast of England. They presumably follow the coasts of France and Spain as they fly south to winter in the Mediterranean or off the west coast of Africa.

Since migration uses much of a bird's energy resources, it may be expected to produce a penalty in higher mortality rates. Between breeding seasons, the Arctic Skua suffers a mortality of about 20 per cent (see section 6.1, chapter 6), which is considerably higher than seabirds like auks, gulls and petrels. What corresponding benefit does the Arctic Skua gain by its annual migration? We can only speculate. Perhaps the way it feeds, as a pirate, or 'kleptoparasite', robbing other seabirds of their food, is less

Figure 2.5 Map showing places where Arctic Skuas ringed on Fair Isle have been recovered.

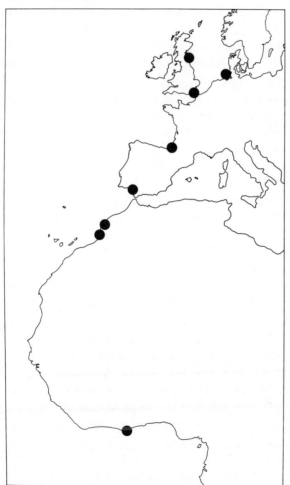

effective outside the breeding season when the large concentrations of birds have dispersed. If larger concentrations of birds can be found in tropical waters in winter, the Arctic Skuas may be forced to go south to find them.

2.5 Distribution of phenotypes: clines in frequencies

For 100 years ornithologists have been particularly interested in the polymorphism consisting of the melanic and non-melanic Arctic Skuas. They have counted numbers of phenotypes in many different breeding colonies. Southern (1943) collected together all the data he could find. He showed that the frequencies of melanics formed a cline lying roughly from south to north. In the far north, on Bear Island (75°N) and Spitzbergen (77–80°N), 99–100 per cent of Arctic Skuas were pale, non-melanics. Further south, at points on about 70°N, 75 per cent were pale. The frequencies of the pale birds decline to about 25 per cent pale in Shetland (60–61°N), Orkney (59°N) and Caithness (58°N). In Finland (60–64°N) only about four per cent are pale. Figure 2.6 shows a map of the cline as if viewed from the North Pole. This map is based on text-figure 5 in Southern's paper with slight modification derived from later data.

The data Southern collected is presented in his table 1. The most useful of these data are the counts of actual numbers of phenotypes in different areas. The remaining data are notes of the phenotypes of isolated birds or impressions of relative numbers. Table 2.7 gives Southern's data where numbers of phenotypes had been counted in different colonies or areas. Figure 2.7 shows a plot of the melanics' frequency in relation to latitude. This plot has the roughly S-shaped form that characterizes frequencies along a cline. The S-shaped curve shown in the figure is not, however, a satisfactory fit: the frequencies vary too much to fit a smooth curve ($\chi_9^2 = 64.24$). In the theory of a diffusion cline maintained by a balance between diffusion and selection (Karlin & Richter-Dyn, 1976) the frequency should change most rapidly in the neutral zone where neither phenotype has any selective advantage. The point where each phenotype is maintained at a frequency of 50 per cent should lie in this zone. On either side of this point each phenotype should increase in frequency towards the region where it has a selective advantage. The 50 per cent point lies roughly on the 67°N line of latitude. If the cline is maintained by a balance between diffusion and selection, the melanic phenotype should be advantageous south of 67°N and the pale phenotype advantageous north of 67°N.

The cline certainly exists. From Orkney in the south to Jan Mayen in the north, the data are set out in table 2.7 in the form of a 2×8 contingency table. In such a table, we may calculate the value of χ^2 for contingency on the hypothesis that the melanic and pale frequencies are the same in all eight areas. This value of χ^2 has seven degrees of freedom. Separate values of χ^2 may be calculated that correspond to each of these seven degrees of freedom. I have analysed the total χ_7^2 into three components: (i) χ_5^2

Figure 2.6 Map of the Northern Hemisphere showing roughly where melanic Arctic Skuas occur at frequencies of 75, 50 and 25 per cent.

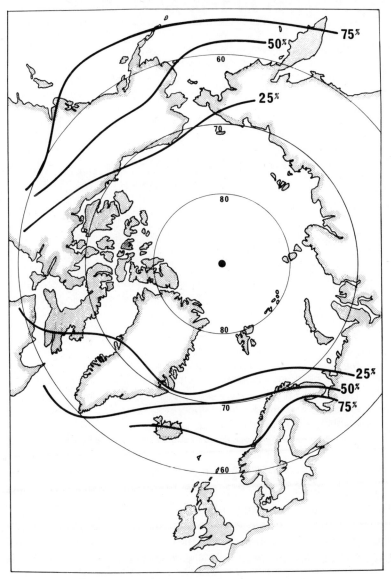

Table 2.7. *Numbers of phenotypes counted at different points along the cline*

Locality	Position	Numbers of phenotypes and percentages of melanics			Date when counts were made
		Melanic	Pale	Total	
Orkney					
Hoy	58°50′N 3°18′W	29 54.72%±6.84	24	53	1943
Papa Westray	59°20′N 2°53′W	21 75.00%±8.18	7	28	1943
Shetland					
Noss	60°10′N 1°00′W	95 74.22%±3.87	33	128	1934–5
Lunna Ness	60°31′N 1°00′W	12 85.71%±9.35	2	14	1941
Hermaness	60°50′N 0°50′W	9 60.00%±12.65	6	15	1902
Iceland					
South	64°01′N 21°12′W	124 72.51%±3.41	47	171	1939–43
North	65°58′N 17°59′W	33 57.89%±6.54	24	57	1939
Jan Mayen	71°00′N 8°00′W	7 11.67+4.14	53	60	1940
Spitzbergen	77–80°N 10–20°E	All pale 0%	—		up to 1931
Edge Land	77°30′N 22°00′E	All pale 0%	—		up to 1931
Norway: Varanger	70°30′N 30°00′E	56 44.09%±4.41	71	127	1937

The positions of the northern and southern localities in Iceland are averages of a number of positions where Arctic Skuas were counted.

corresponding to the five degrees of freedom in the comparison of frequencies in the six southern colonies of Orkney, Shetland and South Iceland; (ii) χ_1^2 for the one degree of freedom in the comparison of the two northern colonies of North Iceland and Jan Mayen; and (iii) χ_1^2 for the one degree of freedom in the comparison of the northern and southern colonies. Table 2.8 shows my analysis of χ^2. The frequencies vary little in the six southern colonies: $\chi_5^2 = 9.0091$ which is in no way significant. But Jan Mayen has a very significantly higher proportion of pale birds than North Iceland; and these two northern colonies both have a very significantly higher proportion of pale birds than the six southern colonies. The probabilities corresponding to the values of χ^2 are negligible: pale birds increase in frequency from Iceland northwards.

Since Southern collected his data, Arctic Skuas have been counted in Shetland (O'Donald, 1960b, 1980c), Iceland (Bengtson & Owen, 1973) and Finland (Hilden, 1971). The new data from Shetland and Iceland may be compared with Southern's earlier data. Thus we can ask: have frequencies changed in these areas of the cline? We cannot ask this question about the cline in Finland because data had not previously been collected there. Table 2.9 shows the complete data from Iceland and my analysis of the data by χ^2. The value of χ_1^2 for the variation in frequency between the two periods of observations was calculated from the 2×2 contingency table,

Period	Melanics	Pales	% Melanic
1939–43	157	71	68.9
1970–71	461	96	82.8

in which the observations from North and South Iceland have been added

Figure 2.7 Plot of the frequencies of melanics along the cline showing the S-shaped curve fitted to the cline.

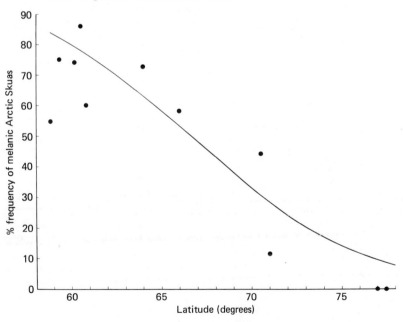

Table 2.8. *Analysis of χ^2 of Southern's data of the cline*

Source of variation in frequency	Value of χ^2	Dfs[a]	Value of P
Difference between southern colonies (Orkney, Shetland and S. Iceland)	9.0091	5	0.109
Difference between northern colonies (N. Iceland and Jan Mayen)	26.7210	1	2.36×10^{-7}
Difference between northern and southern colonies	52.4626	1	—
Total	88.1927	7	—

[a] Dfs = degrees of freedom.

Table 2.9. *Phenotypic frequencies in Iceland*

(i) Numbers of melanic and pale birds in North and South Iceland

Period of observation	North Iceland		South Iceland	
	Melanics	Pales	Melanics	Pales
1939–43[a]	33	24	124	47
1970–71[b]	89	50	372	46

[a] Data of Southern (1943); [b] Data of Bengston & Owen (1973)

(ii) Analysis of χ^2

Source of variation in frequency	Value of χ^2	Dfs	Value of P
Between periods of observations	18.6770	1	1.55×10^{-5}
Between North and South Iceland	42.3661	1	—
Residual (interaction)	1.9117	1	0.167
Total	62.9553	3	—

The value for χ^2_3 was calculated from the overall proportion of melanics, 618/785, and the total numbers observed in each of the four samples. Each sample total removes one degree of freedom; the estimated proportion of melanics removes one more degree of freedom; thus three degrees of freedom are left.

Table 2.10. *Phenotypic frequencies in Shetland*

(i) Numbers of melanic and pale birds on Fair Isle and Shetland

Period of observation	Fair Isle		Shetland	
	Melanic	Pale	Melanic	Pale
1934–31[a]	—	—	107	35
1943–59[b]	162	54	365	129
1973–79[c]	627	157	374	138

[a] Southern (1943); [b] O'Donald (1960b); [c] O'Donald (1980c)

(ii) Analysis of χ^2

Source of variation in frequency	Value of χ^2	Dfs	Value of P
Between periods of observations	2.3386	2	0.311
Between Fair Isle and the rest of Shetland	7.9699	1	0.00476
Residual	0.3094	1	0.578
Total	10.6179	4	0.0312

together. This gives the value

$$\chi_1^2 = 18.6770$$

as shown in the table. Similarly, the 2×2 table for the variation in frequency between North and South Iceland,

Region	Melanics	Pales	% Melanic
North	122	74	62.2
South	496	93	84.2

gives the value

$$\chi_1^2 = 42.3661$$

The third value of χ_1^2 was obtained by subtracting from χ_3^2 the two χ_1^2s for these 2×2 contingency tables. The residual χ_1^2 represents the 'interaction' between the effects of latitude on frequency and the changes in frequencies between the periods of observation. In the interval between 1943 and 1970, there does appear to have been a greater increase in melanics in the South (from 73 to 89%) than in the North (from 58 to 64%). But the value of χ_1^2 does not indicate that this apparent interaction is significant.

There has certainly been a sharp increase in the frequency of melanic birds after 1943, as well as a difference between North and South Iceland. In Iceland, the cline has either not yet reached stability or has started to move. If it has started to move, the neutral point in the cline, where the selective advantage of melanics is lost, must have moved northwards. If the cline has not yet reached stability, the gene for melanism is presumably still spreading northwards: it would do so in the form of a 'wave of advance' forming a moving cline from the front of the wave (Fisher, 1937).

Unfortunately, this evidence of movement in the cline is not supported by the data from Shetland shown in table 2.10. The frequency of melanics has changed very little since 1941. When the data for Fair Isle and the rest of Shetland are combined in a 2×3 contingency table,

Period	Melanics	Pales	% Melanic
1934–41	107	35	75.4
1949–59	527	183	74.2
1973–79	1001	295	77.2

I calculate that

$$\chi_2^2 = 2.3386$$

This shows that the variation in frequency is very much what we should expect by random sampling. The additional data collected after Southern's survey do show, however, that the cline exists in Shetland: Fair Isle, further south, has a higher frequency of melanics than the rest of Shetland. In Southern's original data, differences between colonies in Orkney and Shetland are not significant. The additional data has produced a significant, though only small, difference in melanic frequencies.

The data from Finland (Hilden, 1971) present a completely different picture. Although the samples were taken from five areas ranging from 60°N to 65°N (equivalent to the difference in latitude between Shetland and Northern Iceland), the melanic frequencies vary hardly at all. When the data given in Hilden's paper are combined, we obtain the 2×3 contingency table,

Area	Latitude	Melanics	Pales	% Melanic
I and II	63–65°N	81	5	94.2
III	62°N	97	4	96.0
IV and V	60°N	172	7	96.1

The numbers in Hilden's areas I and II in the north and IV and V in the south have been added together to give sufficient numbers of pale birds in each class for the calculation of χ^2. This has the value

$$\chi_2^2 = 0.05598$$

corresponding to

$$P = 0.756$$

The homogeneity is remarkable over this wide range in latitude; so too is the low frequency of pale birds. In Finland, only 4.4 per cent of birds are pale. Over the same range of latitude from Shetland to Northern Iceland, pales increase from 21 to 36 per cent of the population (Tables 2.9 and 2.10).

The cline has been pushed further north in Finland. Above Finland, the 50 per cent point on the cline in Southern's map lies on about the 70°N line of latitude (figure 2.6). The cline lies roughly on a north-south axis; but there is some variation with longitude as well as latitude. In particular areas like Finland, other factors must partly determine the gene frequencies. It would be futile to speculate what these factors might be. Whatever they are, they presumably increase the selective advantage of melanic Arctic Skuas in Finland. Migration will then raise the frequency of melanism in the areas surrounding Finland, thus pushing the cline further north.

Alternatively, the cline may be explained by the spreading out of the gene for melanism from several original centres. If melanics are generally advantageous, they would spread by gene flow from the centres of their original occurrence, thus producing a moving cline along their wave of advance. But to generate clines moving from south to north, the original centres must all have been at points along the southern edge of the Arctic Skua's range, whence the melanics would have spread northwards. This seems to be a very implausible explanation, for there is no reason to suppose that the same mutation would have occurred only in the same latitudes.

The data that have been collected to the present are just not sufficient to test the two more plausible hypotheses about the evolution of the cline. In Iceland the melanics are moving north. Does this represent a general movement of the cline, caused by a shift northwards in the position of the neutral zone? If so, why has no increase in melanic frequency been observed in Shetland? Or is the change in Iceland only a local effect, caused by some alteration in a balance of selective forces? In a diffusion cline, the rate of change of frequency should be greatest around the neutral zone. At increasing distances from the neutral zone, towards the ends of the S-shaped curve, frequencies become constant. A change in the position of the neutral zone would have little effect at the ends of the cline. Although there is good evidence of a slight cline in Shetland (between Fair Isle and the remainder of the Shetland Isles as shown by the analysis in table 2.10), Orkney shows no significant difference from Shetland in the melanic frequencies; but the Orkney samples are small and were taken a long time ago. From the South of Iceland, through Faeroe to Shetland and Orkney, melanics are roughly at a frequency of 75 per cent. In Orkney and Shetland, the flat end of the S-shaped curve has almost been reached. The neutral zone is about 7° north of Shetland. A shift northwards in its position should have little effect on frequency at the southern end of the cline but might produce rapid changes in populations nearer to the neutral zone. The movement in Iceland and stability in Shetland do not refute the possibility of a general northwards shift taking place in the cline. Extensive counts of phenotypic frequencies along the cline would be necessary to determine the detailed shape of the cline and estimate the position of the neutral zone more accurately. Changes in the cline and the position of the neutral zone could only be determined by repeated sampling at the same colonies over very many years. This would be a life-time's field-work on its own.

There is another hypothesis, already briefly mentioned: the cline is maintained not by diffusion and selection but by a balance between different selective forces that vary with latitude. On Fair Isle, as we shall see in section 6.3, chapter 6, pale birds mature at an earlier average age than melanics. This gives them a selective advantage as a result of their greater chance of surviving to breed. But sexual selection favours the melanics (chapter 6, section 6.3; chapter 8, section 8.4; chapter 10, section 10.5). In theory, sexual selection produced by female mating preference can maintain a balanced polymorphism (see chapter 9, section 9.2): variation in the magnitudes of sexual and natural selection would maintain the variation in gene frequency along the cline. Migration would upset this to some extent. But if the migration only occurs between nearby colonies – colonies in Shetland for example – it will not do much to change frequencies over the greater distances of the cline. On Fair Isle the selective forces might roughly be in balance, though with a large error in their overall estimates. Suppose sexual selection does balance natural selection to maintain the cline. Natural selection would favour the pales and sexual selection the melanics. A cline would be produced if the natural selection favouring pales were more intense at higher latitudes: further north, the point of equilibrium would be reached at a greater frequency of the pales; they would become more common. But if migration and selection combine to produce a diffusion cline, the different selective forces will not be in balance but will favour pales to the north of the neutral zone and melanics to the south. Can one of these theories be refuted? If the overall selection could be measured on either side of the 50% point in the cline, this would corroborate one theory and refute the other. In a diffusion cline, the pale birds should be advantageous in the north and disadvantageous in the south. If sexual selection balances natural selection, selection should have little net effect at either point. These measurements would require another ten years of field-work to carry out. But theoretical arguments can also be used to corroborate or refute one or other of the theories. In the balanced selection theory, only certain selective values and mating preferences are compatible with observed frequencies. Mating preferences and selective values have already been measured in the Fair Isle population. The balanced selection theory can therefore be tested. I develop the theory in section 9.2.3 of chapter 9, estimate the mating preferences in chapter 10, and return to the problem of the evolution of the cline in the concluding chapter of this book.

3

Feeding behaviour and ecology

3.1 Kleptoparasitism by Arctic Skuas in the breeding season and on migration

Arctic Skuas are pirates and predators. On the islands of the North Atlantic, they forage as air pirates, pursuing other seabirds in the air and forcing them to drop their food. This behaviour has become known as 'kleptoparasitism'. Arctic Skuas are said to be 'kleptoparasites'. On arctic tundra in northern Norway, they are predators as well as pirates, feeding on small birds, eggs, rodents, insects, and also on berries like the Crowberry (*Empetrum nigrum*). On Fair Isle, the chicks in particular eat considerable quantities of crowberries.

Auks (Puffins, Razorbills and Guillemots), Kittiwakes and terns (Arctic, Common and Sandwich Terns) are the main victims of the Arctic Skua's piracy. Several detailed studies have been made of the methods Arctic Skuas use in attack and the factors influencing their chances of success (Grant, 1971; Andersson, 1976; Arnason & Grant, 1978; Furness, 1978; Taylor, 1979). Arnason (1978) and Furness & Furness (1980) observed the chances of success in attacks by melanic and pale Arctic Skuas.

Grant (1971) and Arnason & Grant (1978) observed Arctic Skuas in the breeding season attacking Puffins. Taylor (1979) observed them on migration attacking terns. Arnason & Grant (1978) also analysed the factors that influence successful piracy.

Arctic Skuas usually hunt alone (Grant, 1971; Taylor, 1979). They search for victims among flocks of foraging Puffins, terns and Kittiwakes. Taylor remarks that the terns did not attempt to mob the patrolling Arctic Skuas who could thus fly freely in among the flocks of foraging birds. They begin their attacks on Puffins and terns in the same way. They start their chase from above and behind a Puffin or tern, diving to gain speed until they are flying level and below their victim. As they close upon him, they swoop up to attack. They sometimes try to grab the fish dangling from a Puffin's beak. Or they may use their feet to strike the Puffin's back. Puffins are not manoeuvrable flyers. In the chase, Grant (1971) observed that they fly 'more or less straight on'. But Andersson (1976) says: 'Many Puffins responded by evasive flight, involving rapid and erratic changes of direction. Several individuals shook off the skua by a quick turn'. A chase ends when the Puffin either drops its food, or dives into the sea, or reaches its nesting burrow on the cliff. Terns are fast highly manoeuvrable flyers, like the Arctic Skuas themselves. Usually they try to escape by twisting and turning in flight. The skua follows as closely as possible, but rarely manages to strike the tern. An attack ends when the tern drops its food or the skua abandons the chase. If the food is dropped, the skua seizes it either in mid-air or from the surface of the water. Gulls may follow the chases, hoping to seize the food for themselves. This they often manage to do if the food is dropped on land. Skuas seem unable to dive from a low altitude to pick it up. Gulls walking on the land may do so.

Although Arctic Skuas usually hunt alone, they may be more successful in pairs. Table 3.1 shows the numbers of Arctic Skuas hunting in ones, twos and larger groups.

A much higher proportion of single birds attacked terns than Puffins, but the data come from very different populations. Attacking in groups is the more effective tactic. Grant's data were divided into the number of attacks causing food to be dropped or not:

Number of skuas attacking	Number of attacks	
	Food dropped	Food not dropped
1	30	34
2	19	7
3	4	6

This 2×3 contingency table gives the values

$$\chi_2^2 = 5.850, P = 0.0537$$

The greater success of the pair or group of three is nearly significant at the five per cent level of probability.

Taylor found that, in attacks on terns, groups of skuas were very significantly more successful than single birds. However, when several skuas join in attack, not all of them may get food. On average each bird gets less food than when he attacks successfully alone. The chances of success greatly increase when birds attack in pairs. But groups of three or more did not increase their chances compared with pairs. Individuals gain an overall

Table 3.1. *Numbers of attacks on terns and Puffins by one or more Arctic Skuas*

| Number of skuas in attack | Number of attacks observed | |
	On terns[a]	On Puffins[b]
1	343	64
2	56	26
3	18	10
4	4	—
5	2	—

[a] Data from Taylor (1979) of migrating Arctic Skuas in Scotland; [b] Data from Grant (1971) of Arctic Skuas at their breeding grounds in Iceland.

Figure 3.1 Percentage of successful attacks on terns when Arctic Skuas attack in ones, twos, threes, fours or fives. The clear white symbols indicate the percentage of successful attacks by the group; the solid black symbols, the percentage success by each bird.

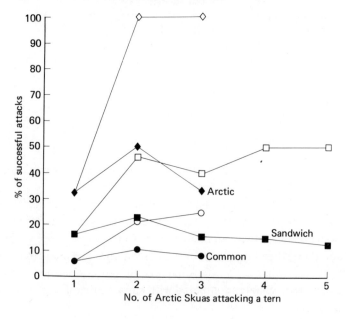

Table 3.2. *Means and standard errors of the times in seconds of attacks by Arctic Skuas*

Species attacked	Attacks by single skuas		Attacks by groups of skuas	
	Successful	Unsuccessful	Successful	Unsuccessful
Arctic Tern[a]	13.1 ± 0.6	11.5 ± 0.8	—	—
Common Tern[a]	6.2 ± 0.3	10.5 ± 0.6	11.3 ± 0.9	22.2 ± 3.7
Sandwich Tern[a]	7.2 ± 0.5	13.1 ± 2.7	15.2 ± 2.2	23.4 ± 4.2
Puffin[b]	12.3 ± 1.6	8.3 ± 0.8	—	—

[a] Data of Taylor (1979); [b] Data of Andersson (1976).

benefit only when they attack in pairs, not when they attack in larger groups. Figure 3.1 shows the results of Taylor's observations (Taylor, 1979). The figure also shows the skuas' greater success in attacks against Arctic Terns compared with their success against Sandwich Terns and Sandwich Terns compared with Common Terns. Arctic Skuas were always successful on the few occasions when they attacked Arctic Terns in pairs. Taylor observes that Common Terns were more agile in flight than Sandwich Terns: they could more easily out-manoeuvre the skuas. The Arctic Terns, although as agile as the Common Terns, were more successfully attacked because they often tried to attack the attacking skuas, dropping their food as they did so. Taylor suggests that the Arctic Terns were migrating south like the skuas, probably from breeding grounds they shared with the skuas. They would already have had to defend their eggs and chicks from predatory skuas. This must certainly be the experience of Arctic Terns on Foula, in Shetland, where Arctic Skuas breed in the middle of the tern colony.

Taylor and Andersson observed the duration of the Arctic Skuas' attacks on terns and Puffins. Taylor timed the attack from the instant a skua started to accelerate towards a tern until either the tern released its food or the skua abandoned the chase. Andersson timed attacks on Puffins. Table 3.2 shows means and standard errors of the durations of attacks on the different species. Andersson's results differ in one respect from Taylor's. Attacks on Puffins were longer when they were successful: attacks on Sandwich and Common Terns were longer when unsuccessful. Successful and unsuccessful attacks on Arctic Terns were not significantly different in duration. Taylor also observed that groups of skuas would attack a tern for about twice the duration of an attack by a single skua. He says

'When only one skua was involved success most often occurred around the time it came into contact with the tern. If the tern did not then release its fish, it was highly unlikely that it would do so later. When attacks were made by groups, some terns released the fish around the time of contact, but others (most often the birds not taken by surprise) only after a prolonged or close-range chase'.

Presumably, a pair's or group's greatly increased chances of success gives advantage to a prolonged chase.

In the attacks on Puffins, unsuccessful chases were shorter than successful ones. If a longer chase may cause a Puffin to drop its food, why does the skua not continue the chase? Andersson gives details of how the chases ended. He timed the following 88 chases of which 19 were successful and 69 unsuccessful:

	Puffin drops food	Puffin does not drop food
Puffin dives to sea	7	41
Puffin flies to cliff	12	28

The skuas seldom continued to chase the Puffins down to sea level. Some skuas did, but the Puffins simply dived below the surface. Puffins have two good methods of escape: diving into the sea, or reaching their burrows. They can thus escape in the middle of the chase. The unsuccessful chases ended sooner than successful ones. Since terns rely on their speed and manoeuvrability to escape, chases end when the skua gives up or the tern drops its food. As we should expect, the unsuccessful chases last longer.

In spite of the high proportion of unsuccessful chases – about 80 per cent in Taylor's and Andersson's studies and 60 per cent in Arnason & Grant's – the Arctic Skua's kleptoparasitism appears to be a very effective method of feeding. Chases only take a few seconds. Andersson found that on average a skua obtained food once every 6 min 21 sec of the time it spent hunting. It is not surprising, therefore, that on Fair Isle skuas hunting for food seem to spend half an hour or less away from their territories. This explains why, after the death of one bird of a mated pair, the survivor can incubate the eggs and rear the chicks successfully: it can collect sufficient food easily and quickly.

Grant (1971) and Arnason & Grant (1978) studied the Arctic Skua's attacks on Puffins in very great detail. The skuas' success rate in forcing Puffins to drop their food depended on the distance from the cliffs and the

Table 3.3. *Outcome of chases at different distances from the cliff and heights above the ground*

(i) Chases at different distances from the cliff

Number of chases by Arctic Skuas

Distance from cliff at the end of the chase (m)	Towards cliff		Away from cliff	
	Fish dropped	Fish not dropped	Fish dropped	Fish not dropped
0–150	140	161	216	179
150–300	41	16	109	32
300–450	12	2	148	80

(ii) Chases ending at different heights above ground

Number of chases by Arctic Skuas

Height at the end of chase (m)	Fish dropped	Fish not dropped	Dropped food taken by		
			Skua	Gull	None
0–50	407	117	192	121	94
50–100	222	141	175	13	34
100–150	76	90	64	0	12

height above the ground at the end of the chase. Table 3.3 shows the data as presented by Arnason & Grant in their paper.

The numbers in the table show some very significant effects. Chases are obviously more successful further away from the cliff and at greater heights above the ground. A skua has a better chance of retrieving food dropped from a greater height. Gulls have a very good chance of getting fish dropped low over land.

The data of table 3.3 (i) is given in the form of a $2 \times 2 \times 3$ contingency table which can be analysed by χ^2. Given the numbers of chases observed at each of the three different distances, the proportion of chases towards the cliffs and the proportion of chases when the Puffins dropped their fish (a total of five independent estimates of parameters for the calculation of expected numbers in each of the twelve classes of the table) I find that

$$\chi_7^2 = 173.4041$$

for the seven remaining degrees of freedom.

In table 3.4, this χ_7^2 is analysed into components corresponding to the effects of distance and direction of chase on the chances of Puffins dropping their fish. These effects might also interact: for example a somewhat greater proportion of Puffins drop their fish when chased at 300–450 m towards the cliff than when they are chased at a similar distance away from the cliff. The interaction of distance with direction corresponds to the residual number of two degrees of freedom when all marginal totals in the $2 \times 2 \times 3$ contingency table are fixed. As table 3.4 shows, this interaction is not significant. The effects of distance and direction by themselves are very significant, however: chases are much more successful at greater distances from the cliff; outward chases are somewhat more successful than chases towards the cliffs. Many more outward chases were observed, as if the skuas knew they had a better chance of success when chasing Puffins out to sea. Puffins with a beak full of fish fly back out to sea if they are prevented from landing beside their burrows. The presence of a nearby observer, for example, will prevent them from landing. They then fly an extended circuit before coming back after a few minutes to make another attempt to land. They must eventually tire if they are forced to make a number of abortive circuits. Heavily loaded Puffins flying out to sea are presumably easier prey to the patrolling skuas.

Table 3.4. *Analysis of χ^2 of chases at different distances and in different directions from the cliff*

Component of variation	Value of χ^2	Dfs	Value of P
Effect of direction of chase on chances of dropping fish	10.3748	1	1.28×10^{-3}
Effect of direction on numbers of chases at different distances	116.1321	2	—
Effect of distance on chances of dropping fish	45.5941	2	—
Interaction of distance and direction on chances of dropping fish	1.3031	2	0.521
Total	173.4041	7	—

There are two directions of chases: towards or away from the cliff. These were observed at each of three distances as shown in table 3.3 (i).

3.2 Selection of Arctic Skuas as a result of kleptoparasitism

Predators must evolve efficient strategies to find and catch their prey. They evolve preferences for prey that provide the most energy for the effort used to catch them. They also learn to discriminate between different prey. They learn what they found good to eat among the prey they caught. This learning can produce selection in favour of different phenotypes in a polymorphic population. Clarke (1962) proposed that the different forms of the snail *Cepaea nemoralis* are selected because predators develop 'scarching images' for the commoner phenotypes that are more often taken as prey. The predators then look for the commoner phenotypes and take them preferentially. Common phenotypes are thus at a selective disadvantage to rare phenotypes. Since the rare phenotypes must increase in frequency as a result of the selection, predators encounter them more often and may switch their searching images. On average the rarer phenotypes, whichever they may be, gain an advantage over the commoner ones: this frequency-dependent selection will maintain a polymorphism of the phenotypes (Clarke & O'Donald, 1964). It has been called 'apostatic selection' because the phenotypes act as signals to the predator indicating suitable or unsuitable prey.

Arnason (1978) made the remarkable suggestion that apostatic selection might maintain the polymorphism of the Arctic Skua, not of course by predation, but by the avoidance reactions of its prey. Generally, apostatic selection in which prey learn to avoid the common phenotype of a predator is impossible because the prey are eaten and cannot learn from their experiences. In this context, Arctic Skuas are pirates, not predators. If their victims were accustomed to be attacked by Arctic Skuas, they might learn the phenotype of their commonest attacker. In areas where melanics are common, they would learn to avoid melanics: pales might be favourably selected. Where pales are common, melanics might be favoured. This is the theory Arnason proposes. He supports it with data apparently showing that in the study area (the south coast of Iceland where pales occur at a frequency of about 11 per cent) the pale phenotype was both the more successful pirate and much the rarer phenotype. Furness & Furness (1980) also observed the success rates of melanic and pale birds in Iceland and several Shetland colonies: they found no significant differences between the phenotypes.

Arnason's own data are of doubtful significance statistically. He used a t-test to comparc the percentages of successful chases by melanic and pale birds. This produced a significant value. But simple χ^2 tests do not confirm Arnason's result. In attacks by single birds, Arnason observed the numbers

	Number of attacks	
	Successful	Unsuccessful
Dark	441	261
Pale	25	6

giving

$$\chi_1^2 = 4.0733$$

which is just significant. But when Yates' correction is applied to obtain a more accurate value, I find

$$\chi_1^2 = 3.340$$

corresponding to

$$P = 0.0676$$

In attacks by pairs of birds, the data

	Number of attacks	
	Successful	Unsuccessful
Dark, dark	194	53
Pale, dark	29	3

give the non-significant value

$$\chi_1^2 = 1.880$$

corresponding to

$$P = 0.170$$

In attacks by three or more than three skuas, pales are still slightly, but not significantly, more successful. The trend, though not significant for any particular size of group, is always for greater success in the presence of pale attackers. Over all sets of data, the trend may be significant. This can be tested very accurately by Fisher's 'Combination of probabilities test', which is based on the fact that

$$\chi_2^2 = -2 \ln(P)$$

I calculated the probabilities with five-figure accuracy from the values of χ_1^2 with Yates' correction. I then applied the transformation shown above and summed the χ_2^2 values to obtain

$$\chi_8^2 = 9.9001$$

corresponding to

$$P = 0.272$$

There is no significant trend in the data. The most important set of data giving the success rates of attacks by single melanic or pale skuas is not quite significant on its own. Thus Arnason's data give only scant support to his hypothesis. Furness & Furness' data give no support at all. And we should require strong and additional evidence before we could accept the theory that apostatic selection maintains the polymorphism.

Even if the pales did have greater success when they were rare and less when they were common, this would not show that frequency-dependent selection had occurred. We must ask: how does the differential success of piracy determine fitness? Neither Arnason nor Furness and Furness consider this question. The answer might simply be: it does not determine fitness at all. Skuas spend so little time feeding, at least in the breeding season, that a reduced chance of successful piracy by a particular phenotype may in no way reduce its chances of survival: even very large differences in success may have negligible effects on survival. No direct and simple correspondence should necessarily exist between the possible differential success of the pirates and their Darwinian fitness. Some relationship might exist if a reduction in the rate of collecting food had so lowered the energy balance of some birds that they were using more energy to collect their food than the food yielded. These birds would then die. If they contained a greater proportion of birds of one phenotype than the population contained as a whole, selection would occur. Even so, the relative fitnesses would depend on what proportion of the population was dying selectively from this cause. A few selective deaths would only produce weak selection. Apostatic selection of predators by the reactions of their prey remains a speculative and implausible hypothesis. No evidence suggests it might be the mechanism that maintains the polymorphism of the Arctic Skua.

3.3 Energetics of feeding behaviour

Arnason & Grant (1978) attempted to measure the energy balance of Arctic Skuas. This would show whether selection could theoretically occur as a result of variation in the success of kleptoparasitic behaviour. If the energy balance declined through the breeding season, this would explain the decline in successful breeding towards the end of the season.

The energy balance is defined as the ratio of the yield in energy to the expenditure of energy. We need to know: how much energy does an Arctic Skua need to use? how much does it obtain from its food? Calculations of these amounts of energy can only be very rough, little more than suggestions of the orders of magnitude. Arnason & Grant calculated as follows. A mean number of 1.08 fish were obtained from a Puffin by each successful chase. The fish in the Puffin's bill weighed 0.655 g (dry weight) on average, corresponding to 3.225 kcal for fish or 3.48 kcal for a successful chase. Arnason found that an adult Arctic Skua weighs 0.4094 ± 0.0029 kg on average. Assuming that the skua's basal metabolic rate, M, is given by the equation

$$\ln M = \ln 78.3 + 0.723 \ln W$$

where M is measured in kcal per day and W is the weight in kg of a non-passerine bird (Lasiewski & Dawson, 1967), it follows that an average Arctic Skua needs 41.05 kcal per day for its basal metabolic functions. From this point onwards the calculation depends on a series of informed guesses of its energy needs while standing or sitting on its territory, while gliding and soaring over the cliffs, and while chasing Puffins. Arnason & Grant assume that about 14 h per day are spent on these activities. They assume that standing, sitting or gliding raises the metabolic rate to $1.35M$ (35% above basal metabolic rate) and powerful flight raises the rate to $15M$. Then if powerful flight occupies only 10 min of the day (a skua making about 50 chases each of 12 s duration), the skua uses 53.3 kcal per day. A 30 per cent rate of success in 50 chases, each success yielding 3.48 kcal, would produce a total of only 52.2 kcal per day. Table 3.3 (ii) shows that skuas secured the food in a proportion of 0.41 chases. This success rate would produce 71.3 kcal per day. Thus an increase in successes from 30 per cent to 40 per cent would convert the energy balance from 0.98 to 1.34: an overall loss of energy would become a gain; selection would favour the more successful individuals. This calculation, which is accurate to no more than an order of magnitude, shows that differences in the rates of successful kleptoparasitism could, at least theoretically, give rise to selection.

An Actic Skua could presumably alter its tactics and hunt more often if it needed more food. It uses 0.0855 kcal in a 12-s chase. Each successful chase produces 3.48 kcal, so only one chase in 40 need be successful. But skuas can hardly spend all their time chasing Puffins. Suppose they must spend about 10 h resting at night using energy at their basal metabolic rate. For the remaining 14 h their metabolic rate is $1.35M$, rising to $15M$ when chasing Puffins. We can then ask: how much of their time must they spend chasing Puffins in order to obtain just enough energy for all their activities throughout a 24-h period? This depends on how successful they are when chasing Puffins. Table 3.5 shows the results of this calculation for different percentages of successful chases. If only 2.57 per cent of all chases were successful, a skua would have to spend all 14 h chasing Puffins. At the higher rates of success, very little time needs to be spent on chases: less successful skuas could easily make up their energy deficit by chasing more Puffins.

Table 3.5. *Time spent chasing Puffins when
different percentages of chases are successful*

Percentage of chases in which skua obtains food successfully	Time (min) spent chasing Puffins to obtain just sufficient energy
40	7.5
30	10.2
20	16.0
10	36.6
5	103
3	372

On the basis of a multivariate analysis of the effects of a number of variables, Arnason & Grant concluded that the rate at which the Puffins brought in food to their burrows was the variable that determined the rate at which the skuas obtained it. This implies that the skuas' rate of feeding increases as the Puffins bring in more food. Arnason & Grant apparently used this result in calculating the energy balance of the skuas during the successive weeks of the breeding season: it rises to a maximum in the third week when the Puffins presumably bring in most food; it falls below 1.0 in the first week and last four weeks. We predicted that the energy balance would fall as breeding success declines towards the end of the breeding season. It falls so far, however, that for at least part of the breeding season, skuas do not appear to be getting enough food. Unfortunately, Arnason and Grant do not explain how they made their calculations. They give no figures except for the path coefficients in their multiple regression analyses. Since the skuas chase only a small proportion of all the Puffins that are bringing food to their burrows, we should expect they would chase a greater proportion of Puffins when smaller numbers were bringing food. Skuas would merely chase more often if their energy balance had dropped below 1.0. But Arnason and Grant state: 'a proportional increase in the number of arriving Puffins resulted in a constant increase in the rate of chasing by Arctic Skuas'. Might there have been some deviation from this simple proportionality? If so, could it have been detected in the data? Arnason & Grant used the method of stepwise multiple regression. As a method it has been strongly criticized. Its logical justification is doubtful: since variables are not selected prior to fitting the model, probabilities to test significance have no meaning. The results derived from the method depend on the combinations of variables and functions of variables chosen for analysis and the sequence in which they are introduced. In Arnason & Grant's analysis, a large residual proportion of the variance was not accounted for by the independent variables they chose to test. The crude approximations in the calculation of the Arctic Skua's energy requirement have been confounded with the relatively small effects of the variables that determine feeding rates and the high level of residual variation. Ecologists have wisely never attempted to estimate the errors in their energy flow calculations. I doubt if the errors in Arnason & Grant's estimates of the energy balance could possibly be calculated. They must be very large. Since Arctic Skuas seldom die in the breeding season, they must be getting the energy they need for themselves and their chicks. Arnason & Grant's figures of the energy balance at different weeks in the breeding season must be under-estimates.

3.4 Territories as feeding areas

Territorial behaviour can be observed in many species of insects and higher vertebrates. Territories are usually assumed to be places where food is collected. But many animals, the Arctic Skuas for example, defend territories although they seldom collect food in them. Territories may be used for sexual display and breeding. They may be places of concealment for the nests or offspring. Sociobiologists (e.g. Wilson, 1975) argue that territoriality gains selective advantage from one or more of these functions, but no direct estimates of the selective advantage have ever been made. The argument is always inductive: territoriality is compared between different species, genera or sub-families in relation to differences in ecology (Lack, 1968). Such arguments may be completely convincing. If widely ranging different groups always show the same association between feeding behaviour and territorial behaviour despite their other differences, the association cannot be ascribed to some other factor of environment or behaviour which the groups all share.

Associations do not indicate the direction of a causal link: feeding behaviour may have caused the evolution of the territorial behaviour; territorial behaviour may have caused the evolution of the feeding behaviour. The standard sociobiological argument goes as follows. Food may be more or less evenly distributed in an organism's environment, or it may be found in clumps or patches. If it is evenly distributed, an organism should defend a territory of a sufficient size to enable it to feed itself and its dependents. Having established a territory round its nest, it can feed freely at a minimum average distance from its nest. If it did not maintain a territory, it would have to compete with other individuals both near and far from its nest. It could not be sure of feeding close to its nest because others might be feeding in the same area: on average it would have to forage further away to get sufficient food. A regular spacing of nests minimizes the time and energy spent travelling in search of food (Brown, 1964; Crook, 1965). But if the food is distributed more or less patchily, a point must be reached at which so few territories would contain food patches that too high a cost would be incurred defending them against other foragers. In the extreme case, when all the food is in a single pile, everyone must come to fight for it. A nest is then ideally placed close to the food source. Selection would favour pairs who breed in a tight, close-knit colony near the food. Of course colonial breeders still maintain small territories immediately surrounding their nests: they must have a place to display, mate, and rear their offspring safely. Thus we conclude: birds whose food is evenly distributed over their environment should be territorial with nests at regular intervals surrounded by territories where they feed; birds whose food is patchily distributed should nest in colonies near where they feed.

How does the Arctic Skua fit into this simple dichotomy? Is it a territorial or colonial species? Obviously these are not separate and distinct alternatives. The Arctic Skua must be classified as both: it certainly nests in colonies; but the pairs are separated, each pair defending a well defined territory. On Fair Isle, territories within the main colony vary from about 1000 m² to 28 000 m² with a mean of about 11 000 m². Nests are about 100 m apart (Davis & O'Donald, 1976b). On Foula the nests are much closer together. In the area just north of the airstrip, some of them are only 10–20 m apart. Even this dense packing produces a loose colony compared to colonies of terns, gulls and gannets. In the tern colony on Foula, nests may be only one or two metres apart. In comparison with terns, skuas are territorial. But their territories may have nothing to do with feeding. As we have seen in previous sections of this chapter, skuas in Shetland are kleptoparasites of terns, Puffins and Kittiwakes. Terns, Puffins and Kittiwakes bring food to their nests on the moors and cliffs. The skuas' breeding grounds are inland on the heather and cotton-grass moors of the islands, nowhere near where they chase their prey. Their territories are too large merely to be places for their nests. In 1963 I suggested that territories might evolve by sexual selection (O'Donald, 1963), for possession of a larger territory might increase the chances of mating. If the males defend the territories, while the females arrive in succession to mate, the males with the larger territories will be more likely to obtain a mate earlier in the breeding season, hence rearing more offspring and gaining a selective advantage. Sexual selection will increase territory size. I shall discuss this theory in detail in section 7.4 of chapter 7.

Although an Arctic Skua's territory is no area in which to feed as a kleptoparasite, it may be a significant area in which to feed as a predator. On the arctic tundra, Arctic Skuas are predators as well as pirates: they breed either as widely separated or isolated pairs or in loose colonies. Andersson & Götmark (1980) studied the tundra-nesting Arctic Skuas at Varanger in north-east Norway. They explain the aim of their study as follows

'The food defendability hypothesis (Brown, 1964) here predicts that solitary pairs to a large extent should forage in defended territories around the nest. Colonial pairs should gather more food far from the nest, for example by kleptoparasitizing the abundant seabirds at the Varanger coast (Norderhaug *et al.*, 1977). Marine food therefore should form a greater proportion of the diet in colonial pairs'.

If Andersson & Götmark's prediction is verified, it will only have established that solitary nesting is associated with one diet and colonial nesting with another. It will not have proved that different feeding habits determined selection in favour of either solitary or colonial nesting. The 'food defendability hypothesis' is an evolutionary hypothesis. If it applies to the Arctic Skua, as Andersson & Götmark assume it may, then Arctic Skuas must be polymorphic: some skuas, being genetically disposed to feed as predators, would have evolved the behaviour of solitary nesting, maintaining large territories in which they can find their prey; others, genetically disposed to kleptoparasitism, would have evolved the behaviour of colonial nesting. Like many sociobiologists, who tend to assume that natural selection can do anything provided an advantage might be gained from some putative trait of behaviour, Andersson and Götmark appear to have ignored the genetical implications of their theory. It clearly requires that genetical variation in feeding behaviour should be closely associated with genetical variation in territorial behaviour: genes for predation must segregate in close linkage with genes for maintaining large feeding territories; genes for kleptoparasitism must segregate in close linkage with genes for nesting in colonies with relatively small terrritories. Only very close linkage could maintain a close association between the adaptive behavioural traits. If the genes were not very closely linked but separated on the chromosomes, genetic recombination would tend to break the association between the traits. Selection would only maintain a weak association between the adaptive behaviours: a somewhat greater proportion of the predatory birds would nest solitarily; a somewhat greater proportion of kleptoparasitic birds would nest colonially. But no close correspondence could be maintained between the behavioural traits. Andersson & Götmark's application of the 'food defendability hypothesis'

Figure 3.2 Maps of Northern Norway where Andersson and Götmark studied the feeding behaviour of pairs of Arctic Skuas.

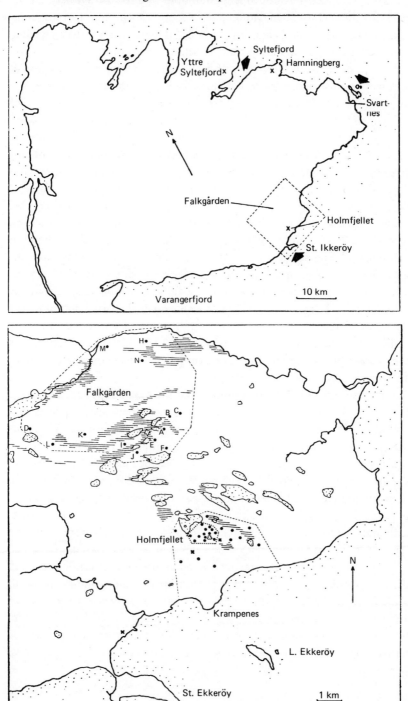

to the Arctic Skua implies that a complex polymorphism of behavioural traits is maintained in Arctic Skua populations by close linkage between the genes for the different behaviours and selection for the adaptive combinations. If so, this would produce the predicted association between feeding as a predator or pirate and nesting as solitary pairs or in colonies.

Are the solitary breeders really as genetically different from the colonial breeders as this theory implies? I think it is very unlikely: it presses the theory of natural selection too far towards being that 'caricature of Darwinism' which Lewontin (1977) sees in much of sociobiology. What of the data? Is diet associated with solitary or colonial nesting? If so, we can then ask: what are the alternative explanations? Andersson & Götmark determined the diets and observed the foraging behaviour of solitary and colonial pairs of skuas in the Varanger peninsula. The maps of figure 3.2 show the localities where the skuas were studied. The colony at Holmfjellet consisted of about 11 pairs in an area of 0.8 km^2 – about 62 000 m^2 for each pair. Outside this colony, pairs were dotted around at about one-tenth of this density. The solitary pairs at Falkgården had an average area of 1.7×10^6 m^2 per pair in one year and 2.4×10^6 m^2 per pair in the next year. These solitary pairs had, potentially at least, 27 times more ground than the colonial pairs. Even the colonial pairs were widely dispersed: a territory of 62 000 m^2 is about two times larger than the largest territories on Fair Isle and six times larger than the average territory. The colony at Syltefjord was somewhat more dense than that at Holmfjellet: 31 pairs occupied 1.5 km^2, an average area of 48 000 m^2 per pair.

Andersson & Götmark determined the diets of pairs from the faecal droppings and pellets. Skuas rest on mounds near their nests. Andersson & Götmark collected the droppings and pellets from their mounds. They thus determined the diets of pairs of birds. The samples were collected at two periods in two successive years. Table 3.6 gives the analysis of prey items in the droppings. The figures are the percentages of droppings that contain each different kind of prey. Solitary pairs generally produced a higher percentage of droppings containing terrestrial food and a lower percentage with marine food than colonial pairs. Data from each pair were used for a series of *t*-tests of the differences between solitary and colonial pairs. Ten pairwise comparisons can be made: four comparisons for the same year and period between colonial pairs at Holmfjellet and solitary pairs at Falkgården and one comparison between colonial pairs at Syltefjord for 26 July 1977 and solitary pairs at Falkgården for 24 July 1977; each of these five comparisons being made twice, once for marine items and once for terrestrial items in the droppings. Andersson and Götmark found that five

Table 3.6. *Prey items in Arctic Skua droppings at Varanger in 1977 and 1978*

| Contents | Colonial pairs at Holmfjellet | | | | Colonial pairs at Syltefjord 1977 | Solitary pairs at Folkgården | | | |
| | 1977 | | 1978 | | | 1977 | | 1978 | |
	7–10 July	19 July	5 July	20 July	26 July	14–16 July	24 July	5 July	18 July
a. Fish	87.3	74.5	77.8	76.4	69.9	77.6	53.7	66.8	61.5
b. Polychaeta/Crustacea	9.0	5.0	17.5	7.3	7.1	1.9	—	2.6	2.5
c. Berries	14.4	31.2	33.5	20.4	30.1	24.4	54.5	29.4	34.5
d. Insects	30.7	44.7	27.8	27.2	34.0	38.8	59.5	29.8	28.5
e. Eggshells	4.2	2.8	1.9	3.7	2.6	5.5	5.8	6.4	5.5
f. Birds	—	1.4	2.4	1.0	3.2	7.5	10.7	18.7	9.5
g. Rodents	0.6	0.7	0.5	0.5	—	5.5	8.3	9.4	3.5
h. Unidentified	1.8	0.7	0.5	0.5	1.9	1.9	1.7	0.8	1.0
i. Marine (a, b)	92.8	78.0	88.2	87.8	75.0	78.0	53.7	67.6	63.0
k. Terrestr. (c–g)	43.4	63.8	51.9	41.4	50.6	56.8	81.8	61.7	60.0
l. No. of pairs	6	6	7	7	6	8	8	8	8
m. No. of droppings	166	141	212	191	156	308	121	235	200

Figures represent % of the group total of droppings (row m) containing the item in question (row a–k). Marine (row i) represents % of droppings which contain items from at least one of categories a and b. Terrestrial (row k) represents % of droppings which contain items from at least one of categories c–g. This table is given as table 1 in Andersson & Götmark (1980).

Table 3.7. *Prey items in Arctic Skua pellets*

Contents	Colonial pairs	Solitary pairs
a. Fish	5	5
b. Berries	7	11
c. Insects	5	7
d. Eggshells	2	3
e. Birds	2	16
f. Rodents	—	8
g. Terrestrial food (b–f)	10	31
No. of pairs	5	4
No. of pellets	10	32

Results were combined for the colonial pairs at Holmfjellet and Syltefjord and for the two years 1977 and 1978.

of the ten comparisons were statistically significant. Although the differences in percentages are not great, solitary and colonial pairs do differ significantly in their diets. As shown in table 3.7, pellets also differed in composition. Solitary pairs produced pellets with more items of terrestrial food. This is not a significant difference:

$$\chi_1^2 = 1.4559 \text{ (with Yates' correction), } P = 0.228$$

At Holmfjellet, skuas from the colony chased the Kittiwakes who were taking food to their own colony at St Ekkeröy. At Syltefjord they chased Guillemots and occasionally Kittiwakes. In the colony at Holmfjellet, skuas were seen to spend only 0.2 per cent of their time foraging within their territory. The solitary pairs were foraging in their territory for 12 per cent of their time.

Andersson & Götmark's data do show that solitary pairs eat rather more terrestrial food than colonial pairs, but not how much more they eat. Differences in feeding behaviour are correlated only to a partial extent with nesting dispersion. A slight correlation is all we should have expected to observe. Genetic differences can determine only a part of the variation in feeding and territorial behaviour. The genes are most unlikely to be maintained in closely linked groups in which, for example, genes for predatory feeding always segregate with genes for solitary nesting. Andersson & Götmark conclude

'To summarize, food and foraging in solitary and colonial arctic skuas differed as predicted by the food defendability hypothesis. We therefore conclude that the two nesting patterns partly represent adaptations to different food situations. Pairs foraging near the nest defend feeding territories, whereas pairs gathering most food at sea nest in colonies, which may be adaptive for several reasons. Anti-predator advantages seem most likely in the present case'.

The 'food defendability hypothesis' would be refuted if the differences in feeding or territoriality had a small genetic component. Since the genetic correlation between these behaviours can only be partial and may well be quite small, a large environmental component in the variation would obscure the correlation completely. Even if the correlation were genetic, this would not show how it had evolved. According to the 'food defendability hypothesis', nesting dispersion evolves as a consequence of feeding behaviour. Equally, feeding behaviour could evolve as a consequence of nesting dispersion. A correlation gives no indication which was the cause and which the effect.

Territorial dispersion of nests probably has several advantages and disadvantages. Andersson & Götmark are probably right in suggesting that colonial breeding may have advantage as a strategy to combat predators. We saw in chapter 2 how Bonxies prey on Arctic Skuas and their chicks. Since Arctic Skuas are so much faster in flight than Bonxies, so easily out-manoeuvre them and attack them so violently, Bonxies must be determined indeed to penetrate far into an Arctic Skua colony. On Fair Isle, however, they soon take chicks that have just started to fly and blundered away from the protection of the colony. The Bonxies watch where the chicks land and drop on them. The most violent attacks by the parent Arctic Skuas seldom succeed in rescuing the chick. Where Bonxies take a large proportion of the chicks, as on Foula, a pair of Arctic Skuas may well raise their fitness by nesting in the middle of the colony. Skuas in the colony must also be dispersed to give the chicks sufficient room to learn to fly. So Arctic Skuas are both territorial and colonial: they need the protection of the colony and a territory in which to display, mate and rear their chicks. As we shall see in sections 7.3 and 7.4, chapter 7, sexual selection may also favour males with larger territories by directly increasing their chances of finding a mate. Territoriality may have evolved as a result of several different selective mechanisms.

But to explain the dispersion of solitary and colonial pairs, is an evolutionary hypothesis even necessary? On Fair Isle, isolated pairs, nesting away from the main colony, are often new birds breeding for the first time. They are often very late breeders, probably having been unable to secure a territory in the midst of the bustle of the colony where most pairs had already laid their eggs. Such isolated pairs often break up after their first year and move into territories in the colony. Birds from the new pairs on Vaasetter and Mopul, who bred very late in 1976, moved into the colony with other mates in 1977 (see map of the nesting grounds in chapter 1, figure 1.1). The pairs on Vaasetter and Mopul had plenty of land to forage in. All Arctic Skuas obtain some food by predation as table 3.6 shows. Isolated pairs may simply have their predatory behaviour re-inforced by the better chances of success in their larger territories. The association of solitary nesting with a more terrestrial diet can easily be explained without the evolutionary maintenance of a complex polymorphism as implied by the 'food defendability hypothesis'.

4

Breeding ecology

4.1 The breeding season

The Arctic Skuas start returning to their breeding grounds on Fair Isle in the third week of April. In the years 1973–75, the first Arctic Skua was seen on the island on 16 April. In 1976 the first bird was seen a day later. Three or four birds have usually arrived by 18 April. Then numbers increase rapidly:

(i) in 1973, 12 birds were seen on 20 April, 80+ on 8 May;
(ii) in 1974, 59 birds were seen on 20 April, 120 on 8 May;
(iii) in 1975, 63 birds were seen on 2 May, 200+ on 15 May.

In 1974, when the most careful and detailed counts of arriving Arctic Skuas were made, a total of 116 pairs eventually bred on Fair Isle. About 25 per cent of the breeding birds had arrived by 30 April, 50 per cent by 8 May. The first arrivals are breeding birds that were colour-ringed in previous seasons. Non-breeding birds arrive much later, forming 'clubs' in late June. As we have already seen (table 2.5, section 2.3, chapter 2), the Bonxies arrive about two or three weeks before the Arctic Skuas. The Bonxies thus gain a great advantage in the competition for territories: a Bonxie may already have established himself on an Arctic Skua's territory before the Arctic Skua returns. The Arctic Skua is forced to seek an unoccupied territory elsewhere. The Bonxies thus drove the Arctic Skuas off their former breeding grounds in the Nature Reserve of Hermaness on Unst, Shetland.

On Fair Isle, the earliest pairs, usually pairs numbered 106 and 118, lay their first egg between 16 and 18 May. By 28 May, half the total number of breeding pairs have laid their eggs. The latest date on which laying has been observed is 18 June. The chick hatched on 14 July. The latest hatching occurred on 22 July. This implies a laying date of about 26 June. Between the earliest and latest laying dates there is a range of 41 days. The early pairs are always well established pairs from previous years. The late pairs are always new pairs, consisting either of new birds breeding for the first time, or of old birds which have changed their mates. No established pair laid after 2 June in 1974: no new pairs laid before 29 May. Established pairs time their breeding to within a day or two of the same date each year. Pair 106, for example, which stayed together throughout the period 1973–79, always laid their eggs between 17 and 19 May, the first egg usually on 17 May.

Hatching dates are much easier to determine than laying dates. Once the first egg has been found, the nest is marked with a cane and then continually checked. Hatching dates have been accurately ascertained for all pairs producing chicks. Pair 106 are usually the first to produce a chick. They have been markedly consistent over the years. In 1973, their first chick hatched on 13 June. Thereafter, until 1978, their first chick hatched on 12 June. In 1979, however, they hatched their first chick on 14 June. The earliest known hatching date is 11 June. This was achieved by pair 118 in 1974. As already mentioned, the latest date is 22 July, giving a range of 41 days in which chicks may be hatched. The timing of breeding may thus vary by about six weeks. Much of this variation arises when birds change their mates. Ignoring second clutches laid after loss of the first, established pairs show a range of hatching dates from 11 June to 5 July – a range of 24 days – with very few dates in July. Within this range, as the dates of pair 106 show, individual pairs are usually very consistent. For example, 15 of all the pairs from 1973 were still together in 1979. Table 4.1 gives their hatching dates analysed to show the small variance within pairs. The mean square within pairs

$$M.S. = 6.353$$

gives a standard deviation

$$s = 2.52 \text{ days}$$

The pairs themselves vary considerably in the consistency of their hatching dates. In addition to pair 106, pairs 116, 205, 225 and 311 have all been very consistent. But pair 219, after breeding later than average in 1974 and 1975, then became a much earlier-breeding pair; the early pair, 302, became a somewhat later pair after 1977.

A chick is fledged when it can fly. The fledging period, which extends from hatching to the first flight, is about 30 days (section 4.3 of this chapter). Since the earliest hatching date has been found to be 11 June, it is not surprising to find that the earliest fledging date is 11 July. The latest recorded date of fledging is 14 August. This was a fledgling that hatched on 12 July. Since some chicks have hatched a week after this date, we may

Table 4.1. *Hatching dates of established pairs from 1974 to 1979*

(i) Individual hatching dates
Pair number

Year	106	115	116	202	205	211	219	225	226	302	310	311	317	331	402
1979	14	—	20	—	15	—	20	17	—	20	22	19	—	18	19
1978	12	19	19	18	13	13	20	18	—	25	23	—	13	18	18
1977	12	20	17	19	13	—	21	15	15	14	20	20	16	19	22
1976	12	—	—	—	—	14	16	—	14	14	—	—	13	27	—
1975	12	22	19	18	15	15	29	16	—	16	24	20	15	25	21
1974	12	17	20	16	14	18	24	16	11	16	20	20	14	25	—

The numbers in this table are dates in June when the pair hatched their first chick. Missing values arise when a pair lost their clutch or failed to hatch a chick, or when the hatching date was not observed.

(ii) Analysis of variance of hatching dates

Source of variation	Sum of squares	Dfs	Mean square	Value of F	Value of P
Between pairs	746.311	14	53.308	8.391	2.17×10^{-9}
Within pairs between years	362.133	57	6.353	—	—

expect they would have fledged on about 21 August. Some fledglings of very late pairs have certainly been observed flying at the end of August, although their actual fledging date was not recorded.

The fledglings soon learn to fly well. They remain with their parents. Presumably a whole family leaves the island together at the end of the breeding season. The immature non-breeding birds are the first to leave. They arrive late in June and leave in July. The first breeding birds leave with their chicks in August. Most of the Arctic Skuas have left by about 7 September. Counts show that by this date only about 10 birds are left:

(i) in 1974, 10 birds were left on 9 September and the last bird was seen on shore on 19 September;
(ii) in 1976, last bird was seen on shore on 10 September;
(iii) in 1977, last bird was seen on shore on 10 September;
(iv) in 1979, 8 birds were left on 7 September.

These are records of Arctic Skuas present on the island. Presumably they were Fair Isle breeding birds. Arctic Skuas can be seen on migration off shore until 28–29 September. The latest record was an Arctic Skua seen off shore on 3 October.

If the last chicks and their parents are leaving on about 7 September, the chicks, having fledged on about 14 August, would be leaving about three weeks after fledging. This must be the time they need to become sufficiently proficient fliers to set off on migration. Two weeks after fledging, they certainly seem to be flying almost as well as their parents. They may need yet more time to learn to become effective pirates and feed themselves. They probably remain dependent on their parents for some time after setting off on migration.

The Arctic Skua's breeding season on Fair Isle starts when the first birds arrive back on their breeding grounds on 16 April. It ends when the last birds leave on 10 September. One hundred and forty-seven days have passed from the first arrival to the last departure. Figure 4.1 shows how this period is divided into a period before eggs are laid, a period when eggs may be found, a period with unfledged chicks and a period after fledging.

4.2 Egg laying and incubation

The date of egg laying is perhaps the best single measure of breeding date. It is determined largely by the physiological state of the female. Other dates, such as hatching or fledging, may be influenced by other factors such as the continuity of incubation or the feeding abilities of the parent birds. Laying dates, particularly the date when the first egg in a clutch is laid, are difficult to observe. Since the nest is no more than a slight scrape in the grass or heather, it cannot be found before an egg has been laid in it. To determine the dates when the first eggs are laid – for example the dates given in the first section of this chapter – the early pairs must be kept under daily observation and their territories thoroughly searched. Fortunately, the birds themselves give a clue: they only start to feign injury and defend their territories against human intruders when they have an egg or are just about to lay (see section 7.2, chapter 7). The first desultory swoops

Figure 4.1 The breeding cycle of the Arctic Skua on Fair Isle.

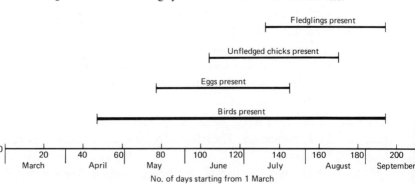

or attempts to feign injury are an indication that an egg has just been laid or will be laid very soon. Then an intensive search can be made to find the egg and nest. Once the first egg has been found and the nest marked, the date when the second egg is laid will be noted during one of the daily checks of each nest in the colony. Laying dates were only observed intensively in the years 1973–75 when John Davis was working as a full-time research assistant to the Arctic Skua project. All the data on laying dates are derived from these years.

4.2.1 Clutch size

Arctic Skuas normally lay two eggs. We found only three clutches with three eggs. Clutches of one egg are usually produced by late breeders, often new pairs. We observed the following numbers of clutches with 1, 2 or 3 eggs:

	1 egg	2 eggs	3 eggs
numbers observed	108	725	3
percentage	13	87	0.4

Numbers of clutches produced at different dates and by new or old pairs are shown in table 4.2. The analysis of χ^2 shows that clutches of one egg are produced very significantly later than clutches of two eggs. Since new pairs breed later on average than old pairs, they tend to produce more of the one-egg clutches. This effect is also very significant statistically. But some residual heterogeneity is left after both these effects have been eliminated: it is determined by a difference between the old and new pairs in the relative frequencies of one-egg clutches that each of these pairs produce at different periods in the breeding season. New pairs produce a relatively smaller proportion of one-egg clutches at the beginning of the season and a relatively greater proportion at the end. This statistical 'interaction' between breeding date and type of pair is only just significant. It may contribute slightly to the large differences in the fledging success of pairs breeding at different times in the breeding season. On these differences in fledging success, the sexual selection of the phenotypes depends (chapters 8 and 10).

4.2.2 Incubation period, order of laying the eggs and breeding experience

The incubation period of the Arctic Skua is roughly 26 days. It is seldom less than 25 or greater than 27 days. Table 4.3 shows the actual distributions of incubation periods for each of the years 1973, 1974 and 1975. The median period is always 26 days, but the means vary slightly between years. This variation in the means is, indeed, almost significant at $P = 0.06$. In 1973, when John Davis and I started the Arctic Skua project on Fair Isle, we did not know where the birds sited their territories or where they were likely to make their nests. We may have missed their exact laying dates. Errors in observation, which must consist of failing to find an egg on the day it was laid, can only reduce the apparent incubation period, not increase it. Yet in 1973, the incubation period is slightly greater than in the subsequent years. This discrepancy is difficult to explain, though seasonal variations may contribute to variation in incubation period: a very cold season may slightly increase the time of development as the eggs chill more rapidly when left unincubated.

Once the first egg has been laid and the nest marked, the laying date of the second egg is found to the nearest day during the daily nest checks.

Table 4.2. *Clutch size of Arctic Skuas in relation to breeding date*

(i) Numbers of clutches of old and new pairs

Date in weekly intervals	Old pairs		New pairs	
	1 egg	2 eggs	1 egg	2 eggs
10–16 June	—	82	—	3
17–23 June	8	165	2	31
24–30 June	9	94	9	83
1–7 July	7	14	14	40
8–14 July	—	3	7	19
15–21 July	—	1	4	6
Totals	24	359	36	182

In this table, 'old pairs' consist of birds who have bred together in previous years as a pair on Fair Isle. 'Date' is the date when the first egg hatched.

(ii) Analysis of χ^2

Component of variation	Value of χ^2	Dfs	Value of P
Difference in dates of new and old pairs	169.6627	2	—
Difference in dates of 1- and 2-egg clutches	54.7422	2	—
Difference of new and old pairs in proportions of 1- and 2-egg clutches	16.2335	1	5.60×10^{-5}
Residual heterogeneity	6.2039	2	0.0450

To avoid using very small expected numbers in calculating the values of χ^2 in this table, the numbers observed in the first two and last three weeks of the breeding season were added together, leaving only three intervals in breeding dates. The data were thus analysed as a $2 \times 2 \times 3$ contingency table, with a total of seven degrees of freedom after allowing for the three totals of the three intervals in dates, the proportion of new pairs and the proportion of one-egg clutches.

Table 4.3. *The incubation period of the Arctic Skua*

(i) Data of incubation periods in different years

| Incubation period (days) | Numbers observed in different years | | | |
	1973	1974	1975	All years
24	1	7	7	15
25	6	33	40	79
26	9	34	49	92
27	4	9	11	24
28	1	1	1	3
29	1	0	0	1
Total	22	84	108	214
Median	26	26	26	26
Mean	26.05	25.57	25.62	25.64

(ii) Analysis of variance

Source of variation	Sum of squares	Dfs	Mean square	Value of F	Value of P
Between years	4.0481	2	2.0241	2.7921	0.0636
Within years	152.9612	211	0.7249	—	—

Table 4.4. *Incubation periods of first and second eggs*

(i) Data of incubation periods

| Incubation period (days) | Parity | |
	1st egg	2nd egg
24	12	3
25	38	41
26	53	39
27	17	7
28	3	—
29	—	1
Total	123	91
Median	26	26
Mean	25.68	25.59

(ii) Analysis of variance

Source of variation	Sum of squares	Dfs	Mean square	Value of F	Value of P
Parity (1st or 2nd egg)	0.4191	1	0.4191	0.5674	0.452
Within 1st and 2nd eggs	156.5902	212	0.7386	—	—

Since hatching dates are also found during the daily checks, the incubation periods of second eggs are accurate to a day. Errors should only affect the incubation periods of first eggs which may not have been found on the day they were laid. Since Arctic Skuas start incubation immediately the first egg is laid, the true incubation periods of both first and second eggs should be the same. Table 4.4 gives the data. There are more data for incubation periods of first eggs, since the incubation periods for single eggs have been included with the incubation periods of first eggs in clutches of two eggs. The data in the table and the analysis of variance show that first and second eggs are incubated for almost exactly the same periods: the means are almost identical.

Some birds are clearly more ready to incubate than others. This becomes apparent when trying to catch them. They can only be caught while incubating, and some are most reluctant to incubate when faced by funnel traps over their nests or clap-nets alongside. Some birds simply refuse to enter a funnel trap, but may be taken by clap-net worked from a hide. Others refuse to approach the nest when any trap or net is present. Three birds remained uncaught for several years despite many days spent accustoming them to hides and nets, and hours spent in attempts to catch them. They may have been broody but too easily distracted, or just reluctant to incubate, leaving the incubation to their partner. Birds of other pairs may both show extreme readiness to incubate: both may be caught in a funnel trap within a few minutes of each other. Pairs probably differ in the continuity of the incubation they provide: some may incubate almost continuously; others may leave the eggs without incubation for longer or shorter intervals. Incubation periods may thus differ between pairs. In particular, incubation may depend on experience: new pairs may be less efficient or effective than old pairs with previous experience of breeding together. The data in table 4.5 show that new and older pairs have almost the same incubation periods. The difference between them is very small and obviously not significant.

4.2.3 Incubation period and phenotype

We have seen, in the data of tables 4.4 and 4.5, that incubation period is the same for first and second eggs and new and old pairs. Incubation period varies by little more than one day on either side of the median value of 26 days. About 91 per cent of the eggs that are going to hatch do so after 25, 26 or 27 days incubation. Most of the rest take 24 days. Around an overall mean of

$$\bar{x} = 25.64 \text{ days}$$

the standard deviation in incubation period is

$s = 0.859$ days

Hatching date corresponds very closely with laying date. This fact is crucial to my analysis of sexual selection in the Arctic Skua. As we shall see later in this book (chapters 8 and 10), my analysis depends on knowing the relative breeding dates of different pairs. Although the date of laying the first egg is probably the most satisfactory measure of breeding date, it can be ascertained for at most about half the total number of pairs and only after a great deal of hard work at the beginning of the breeding season. But hatching date can easily be ascertained for all pairs that succeeded in hatching an egg. Owing to the requirements of University examinations, I was usually unable to arrive on Fair Isle before June. On arrival I could easily find the nests and then start nest checking from 11 June onwards in order to observe the hatching dates. But without the full-time assistance of John Davis, who left the project in 1976, no laying dates could be ascertained. Thus I have used hatching date as a relative measure of a pair's breeding date. According to this measure, the non-melanic pale males take longer to find a mate than the melanic males. Females prefer to mate with melanic males (chapter 10). Melanic males are chosen as mates earlier in the breeding season than pales. This conclusion depends on the assumption

Table 4.5. *Incubation periods of new and old pairs*

(i) Data of incubation periods

Incubation period	Experience of pair	
	New pairs	Old pairs
24	3	11
25	11	62
26	26	57
27	3	17
28	0	2
Total	43	149
Median	26	26
Mean	25.67	25.58

(ii) Analysis of variance

Source of variation	Sum of squares	Dfs	Mean square	Value of F	Value of P
Between new and old pairs	0.3155	1	0.3155	0.4765	0.491
Within pairs	125.8043	190	0.6621	—	—

that the hatching dates (of the first egg) are valid measures of the breeding dates of the melanic and pale phenotypes. But if the chicks differed in their rates of development in the egg, the phenotypes might differ in their incubation periods. This would have some slight effect on the hatching dates of the parent. Suppose, for example, that pale chicks developed rather more slowly and hatched a day later than intermediate or dark chicks. Matings of pale adults, producing pale chicks, would hatch their chicks a day later on account of their chicks' longer incubation period. This would then account for some part – a small part – of the breeding dates of the pale males.

Do the phenotypes differ in their incubation periods? The answer to this question is provided by the data given in table 4.6. The means are very similar with no significant difference between them. The different years

Table 4.6. *Incubation periods of the phenotypes*

(i) Data of incubation periods

Numbers of phenotypes in different years

Incubation period (days)	1973			1974			1975		
	D	I	P	D	I	P	D	I	P
24	1	—	—	3	4	—	3	4	—
25	1	5	—	7	21	4	3	30	5
26	3	5	—	3	19	7	9	36	4
27	1	2	—	1	5	1	1	7	1
28	—	1	—	—	1	—	—	1	—

(ii) Medians and means for all three years

	Phenotype		
	D	I	P
No. observed	36	141	22
Median	25–26	26	26
Mean	25.39	25.63	25.68

(iii) Analysis of variance

Source of variation	Sum of squares	Dfs	Mean square	Value of F	Value of P
Between phenotypes	1.8792	2	0.9396	1.3936	0.251
Within phenotypes	132.1510	195	0.6742	—	—

The symbols D, I and P refer to dark, intermediate and pale phenotypes.

show very similar distributions. We can conclude that the phenotypes are incubated for very similar periods. Even if the small differences in the means were general and not the consequences of sampling, they would have negligible effects on the parental phenotypes' relative hatching dates. Hatching date can therefore be taken as a very good measure of breeding date:

(i) incubation usually takes 25 or 26 days;

(ii) the order in which the eggs are laid has no effect on the incubation period;

(iii) the phenotypes have similar incubation periods.

4.3 Fledging the chicks

A chick's rate of development is fairly constant while the egg is being incubated. After hatching the chick develops at a rate that depends largely on how well it is fed. The time to fledging will obviously vary more than the time to hatching. The date of actual fledging is not likely to be ascertained as accurately as the date of hatching. Some chicks, after fledging, may be more reluctant to fly than others. They may still retain the instinct to crouch at an intruder's approach. They will not be put into the air so easily. So the moment when a chick can fly – when it is fledged – is less determinate than the moment when a chick breaks free from the egg shell – when it is hatched. Measurement of the fledging period is necessarily less precise than measurement of the incubation period. This uncertainty in measurement will also increase the variability in fledging dates.

After fledging, the chicks are still dependent on their parents for food. As we saw in section 4.1, the last birds leave Fair Isle about three weeks after the last chicks are fledged. The chicks probably remain dependent on their parents at least until they leave, and probably after they have left, eventually becoming independent while on migration. There is no evidence on this point. To ascertain the Arctic Skua's age at independence, it would be necessary to follow an identifiable family group as they set off on migration: this would be virtually impossible.

Our data on the fledging period have been obtained during visits to territories after the chicks have hatched. We aim to ring the chicks at about 20 days old. We then know how many chicks have survived the early hazards of life. The main cause of early death is a storm just after the chicks have hatched. We have found drowned chicks near the nests. After the survivors have been ringed, we continue to visit the territories to observe when the chicks fly for the first time. Just before they are ready to fly, they can usually be observed standing and walking rather than crouching motionless like younger chicks. The first flights are very laboured. The chicks can only fly a few metres. They can still be caught at this stage. But after only two or three days they are flying well. After two weeks they seem as agile in the air as their parents. Even so, they still keep mainly within the general area of their territories. Of course they blunder into other skuas' territories. If the other skua is a Bonxie, this is usually fatal. Most of the predation of Arctic Skua chicks occurs just after they have learnt to fly and are easy prey to the powerful Bonxies (see section 2.3 of chapter 2). After their chicks have fledged, Arctic Skuas cease to defend their territories so vigorously, except against Bonxies. They allow other pairs' chicks to fly into their territories without taking offence. By this time, too, they have ceased to attack human intruders.

Table 4.7 gives our data of fledging periods for the years 1973–75. In later years, without the full-time assistance of John Davis, I made no attempt to

Table 4.7. *The fledging period of the Arctic Skua*

(i) Data of fledging periods in different years

Fledging period (days)	Numbers observed in different years			
	1973	1974	1975	All years
26	—	—	1	1
27	2	1	1	4
28	5	5	4	14
29	4	7	9	20
30	13	22	10	45
31	11	24	8	43
32	10	18	10	38
33	6	5	2	13
34	5	13	—	18
35	3	4	—	7
36	3	3	—	6
37	—	2	—	2
38	3	—	—	3
Total	65	104	45	214
Median	31	31	30	31
Mean	31.63	31.53	30.22	31.285

(ii) Analysis of variance

Source of variation	Sum of squares	Dfs	Mean square	Value of F	Value of P
Between years	64.782	2	32.391	6.898	0.00125
Within years	990.830	211	4.696	—	—

ascertain fledging periods: I was concentrating on observing the survival rates, changes of mate and breeding dates to obtain further data for the estimation of the parameters of sexual and natural selection. The data in table 4.7 clearly show that the chicks developed slightly faster in 1975, fledging about a day sooner than in the previous years. The difference between 1975 and 1973–74, though small, is highly significant; it accounts for almost the whole of the sum of squares between years in the table of the analysis of variance. The shorter fledging period in 1975 is not surprising: 1975 was extraordinarily fine and dry in Scotland. During the period when the chicks were hatching, John Davis and I spent three weeks surveying and ringing the Arctic Skuas on Foula. The Warden of the Fair Isle Bird Observatory, Roger Broad, carried on our work on Fair Isle. For these three weeks when most of the chicks were hatching and starting development, the sun shone continually from a cloudless sky of endless visibility. We became very sun-burnt wearing only shorts even on Foula. This was the hottest weather I experienced in Shetland. In these conditions, so different from the usual cold and rain, an enhanced rate of development was only to be expected. If so, the fledging period would also have been somewhat shorter, as indeed we found.

Fledging obviously takes place over a much wider range of time than incubation. The standard deviation of the fledging period is much greater than that of the incubation period. We can estimate the standard deviation by taking the square root of the mean square within years shown in table 4.7. This gives the estimate

$s = 2.17$ days

for fledging compared with the estimate

$s = 0.86$ days

for incubation. Since a chick's rate of development determines when it fledges, fledging must be influenced by many extrinsic factors, particularly feeding rate, which thus increase the variability of the fledging period.

Fledging success sharply declines towards the end of the breeding season (see table 8.6, section 8.5, chapter 8). If this decline occurs because late in the season the parents are less successful in finding sufficient food for their chicks, we should expect to observe a reduced rate of development and hence an increased fledging period of later chicks. To test this hypothesis, I have calculated the regressions of fledging period on breeding date (measured, as usual, by the date of hatching of the first egg). The results of my calculations are shown in table 4.8. Both the overall regression of fledging period on breeding date and the correlation are nearly significant:

$P=0.082$, with no heterogeneity between years. But the regression is the opposite of our expectations! The earlier pairs have the slightly longer fledging period. How might this be explained if the correlation were real and not an accident of sampling? Rather than attempt to answer this question, I would prefer to accept that the null hypothesis – that there is no correlation – has not been refuted, and assume that nothing has been found that presently requires explanation.

The parents may vary in their abilities to care for and feed their chicks. Such variation would also influence rate of development. If so, those pairs who have previous breeding experience together may be better parents than new pairs. Many new pairs consist of birds breeding for the first time. They are much less successful in fledging chicks than experienced breeding birds (see table A3 of appendix A). Their relative lack of success may indicate

Table 4.8. *Regression analysis of fledging period and breeding date*

(i) Regression and correlation coefficients

Year	Regression coefficient	Correlation coefficient
1973	−0.0411	−0.122
1974	−0.0584	−0.180
1975	−0.0039	−0.021
All years	−0.0363	−0.119

(ii) Regression analysis of variance all years combined

Source of variation	Sum of squares	Dfs	Mean square	Value of F	Value of P
Regression	14.957	1	14.957	3.0470	0.0823
Residual	1040.655	212	4.909	—	—

(iii) Test of homogeneity of regressions in different years

Source of variation	SS_x	SP	SS_y	Dfs	Residual SS_y	Dfs	Regression SS
1973	3764.462	−154.692	429.138	64	422.782	63	6.3567
1974	4275.990	−249.721	447.914	103	433.330	102	14.5839
1975	3257.911	−12.556	113.778	44	113.729	43	0.0484
Total	11298.363	−416.969	990.830	211	975.442		

Residual of individual regressions   969.841   208

Difference for homogeneity   5.601   2

Test of homogeneity: $F_{2,208} = (5.601/2)/(969.84/208)$
$$= 0.6006$$
$$P = 0.549$$

In this table SS_x is the sum of squares of breeding dates, SP the sum of products of breeding dates and fledging periods, and SS_y the sum of squares of fledging periods.

they are not finding enough food for their chicks. Those chicks that do survive may develop more slowly with a longer fledging period than the chicks of experienced pairs.

The test of this hypothesis is given in table 4.9. The breeding pairs of 1974 and 1975 can be separated into new pairs and old, since the adult birds were first ringed in 1973. Since the years 1974 and 1975 differ significantly in fledging period, the data have been classified separately for each year. To some extent, the difference between years will affect any difference that may exist between the fledging periods of the different pairs. The difference

Table 4.9. *Fledging periods of new and old pairs in different years*

(i) Data of fledging periods

Year and experience of pair (new or old)

Fledging period	1974		1975		Both years	
	New	Old	New	Old	New	Old
26	—	—	—	1	—	1
27	—	1	—	1	—	2
28	3	2	1	3	4	5
29	3	4	3	6	6	10
30	8	14	5	5	13	19
31	12	12	5	3	17	15
32	6	12	1	9	7	21
33	2	3	1	1	3	4
34	3	10	—	—	3	10
35	1	3	—	—	1	3
36	—	3	—	—	—	3
37	1	1	—	—	1	1
Total	39	65	16	29	55	94
Mean	31.15	31.75	30.31	30.17	30.91	31.27

(ii) Analysis of variance

Source of variation	Sum of squares	Dfs	Mean square	Value of F	Value of P
Between 1974 and 75	42.5325	1	42.5325	11.158	0.00107
Between new and old pairs	1.5327	1	1.5327	0.4021	0.527
Interaction	3.9688	1	3.9688	1.0421	0.309
Residual	552.714	145	3.8118	—	—

In this orthogonal analysis, the effect of the difference between years in no way contributes to any difference that may exist between old and new pairs. The interaction is a measure of whether the difference between old and new pairs differs from one year to the next. The year 1973 supplies no data on the experience of the pairs since that was the first year of the project.

between years and the effect of experience of pairs must be tested independently of each other. The analysis of variance must be 'orthogonal'. This means that the sum of squares for the different years must be independent of the sum of squares for new and old pairs. An orthogonal analysis is simple to derive when equal numbers have been observed in each of the classes. Programmable calculators have standard programmes to produce these analyses. In experimental work, experiments are usually designed so that the numbers of observations in each class are equal. When numbers are necessarily unequal, as in randomly collected values, an orthogonal analysis is much more difficult to derive. Appendix B derives the orthogonal analysis of data with $2 \times r$ classes of observations. The

Table 4.10. *Fledging periods of first and second chicks in different years*

(i) Data of fledging periods

	Years and order of hatching of the chicks					
Fledging period (days)	1973–74		1975		All years	
	1st chick	2nd chick	1st chick	2nd chick	1st chick	2nd chick
26	—	—	1	—	1	—
27	2	1	1	—	3	1
28	6	4	1	3	7	7
29	9	2	7	2	16	4
30	16	19	4	6	20	25
31	17	18	4	4	21	22
32	18	10	7	3	25	13
33	5	6	1	1	6	7
34	8	10	—	—	8	10
35	3	4	—	—	3	4
36	3	3	—	—	3	3
37	1	1	—	—	1	1
38	1	2	—	—	1	2
Total	89	80	26	19	115	99
Mean	31.36	31.80	30.19	30.26	31.10	31.51

(ii) Analysis of variance

Source of variation	Sum of squares	Dfs	Mean square	Value of F	Value of P
Between years	63.6777	1	63.6777	13.603	2.88×10^{-4}
Between 1st and 2nd chicks	2.2766	1	2.2766	0.4863	0.486
Interaction	1.1896	1	1.1896	0.2541	0.615
Residual	983.0171	210	4.6810	—	—

analysis of variance shown in table 4.9 has been calculated using the formulae of table 2B (appendix B). This shows that the rather small difference between the fledging periods of old and new pairs is in fact less than what we should expect to arise by random sampling ($F < 1.0$). It is obviously not significant. But it is the opposite of what we had predicted might arise: new pairs have slightly shorter fledging periods than old pairs. Since much larger differences could easily arise by chance, nothing need be made of this fact. Breeding experience seems to have little or no effect on fledging period.

The order in which the eggs were laid may have an effect. The first chick to hatch has about two days' start in development over the second. The first chick is almost always larger. If chicks compete for food, the first and larger chick should have the advantage, getting more of the food and developing more rapidly. If so, its fledging period should be shorter. The analysis of the data of table 4.10 tests this hypothesis. As in table 4.9, the data of 1975 are separated from the data of 1973–74 because the fledging period was significantly shorter in 1975. An orthogonal analysis of variance is again used to eliminate the effect of the difference between years from the comparison of first and second chicks. The results are not significant. First and second chicks develop at almost exactly the same rates and fledge after the same average period.

A day-old chick beside the chipped egg of its sibling which will hatch in about 24 hours.

Table 4.11. *Fledging periods of phenotypes in different years*

(i) Data of fledging periods
Years and phenotypes

Fledging period (days)	1973			1974			1975		
	D	I	P	D	I	P	D	I	P
27	1	1	—	1	—	1	—	—	1
28	1	4	—	2	2	2	1	3	—
29	1	3	—	—	5	1	1	6	2
30	1	9	3	—	21	5	1	6	2
31	3	6	2	6	13	2	2	5	1
32	2	7	1	1	15	1	3	7	—
33	1	5	—	3	1	—	1	—	—
34	2	1	2	3	10	1	—	—	—
35	1	2	—	2	1	—	—	—	—
36	—	3	—	1	2	—	—	—	—
37	—	—	—	—	2	—	—	—	—
38	1	2	—	—	—	—	—	—	—
Totals	14	43	8	19	72	13	9	27	6

D represents dark, I intermediate and P pale.

(ii) Mean values for 1973–74 and 1975

	Dark	Intermediate	Pale
1973–74	31.91	31.54	31.19
1975	30.89	30.26	29.33

(iii) Analysis of variance

Source of variation	Sum of squares	Dfs	Mean square	Value of F	Value of P
Between years	43.0519	1	43.0519	9.2906	0.00261
Between phenotypes	14.6096	2	7.3048	1.5764	0.209
Interaction	2.0101	2	1.0050	0.2169	0.805
Residual	949.9467	205	4.6339	—	—

The interaction measures how the difference between years varies between each of the three phenotypes.

In view of the small difference in the means of our observations, often less than expected on random sampling, we can conclude that the breeding experience of the pairs and the order in which the eggs were laid and chicks hatched can have virtually no effect on the Arctic Skua's fledging period.

4.3.1 *Fledging period and phenotype*
Up to this point in section 4.3, I have examined whether the order of laying the eggs or the experience of the pairs influences the Arctic Skua's

fledging period. Such extrinsic factors might give rise to some of the variance in fledging period which, as we have already seen, is much more variable than incubation period. But on examination, we found they produced no significant effects.

A chick's genotype is a purely intrinsic factor which might determine rate of development and fledging period. If dark, intermediate, and pale phenotypes develop at different rates and fledge after shorter or longer periods, this might also determine a component of the natural selection of the phenotypes (analysed in section 6.3 of chapter 6); the slower developers would be exposed longer to predators and set off later on migration.

The data of the fledging periods of each of the phenotypes are shown in table 4.11. As in previous analyses of fledging periods, I have classified the data by year to allow for the significantly shorter fledging period in 1975. The mean periods are shown for the years 1973–74 and 1975. These means form a 2×3 table of values. Their contributions to the sums of squares in the analysis of variance have been calculated by the formulae of table B2 (appendix B). The pale birds had the shortest fledging period and the dark birds the longest. This difference is consistent over the years. But the numbers of pale and dark birds in the samples are too small to detect a significant effect at the magnitude of the observed differences. The analysis of variance shows that the phenotypes do not differ significantly in their fledging periods.

To summarize: we have found no evidence of intrinsic or extrinsic factors that determine any significant part of the variation in the Arctic Skua's fledging period. Its fledging period can vary from 26 to 38 days around the values: median = 31 days; mean = 31.29 days.

5

Genetics

5.1 Genetic analysis of matings between phenotypes

The first step in studying any polymorphism is to determine the genetics of the different phenotypes. At first sight, the polymorphism of the Arctic Skua seems to depend on a simple genetic mechanism. As adults, the pale, non-melanic birds are completely distinct from the dark and intermediate melanics. Dark and intermediate birds show a continuous range of phenotypic expression with no clearly definable difference separating darks from intermediates. Matings of pale × pale birds have only produced pale offspring. We may thus infer: (i) that pale birds are homozygous; and (ii) that melanism is semi-dominant to pale. If only two alleles determine the phenotypes, pales must be homozygous for the pale allele, while darks and intermediates are either homozygous or heterozygous for the alternative, melanic allele. Presumably the melanic homozygotes would tend to be darker than the heterozygotes.

5.1.1 Chicks produced by matings

A difficulty is encountered when classifying the chicks produced by matings of different phenotypes. The phenotypes of the chicks show no absolutely clear differences. All gradations can be observed, from pale chicks with a large patch of white feathers on their belly, to those with just a few white feathers, then to those with some darkening at the tips of the mainly white feathers, through intermediates with bands of dark pigment, and so to darks with completely dark belly feathers. Although there is this continuous gradation from pale to dark, the great majority of birds are easily classified by phenotype. (Section 1.4 of chapter 1 gives detailed descriptions of the phenotypes.) The distinction between pale and intermediate is much less in chicks than in adults. Intermediate chicks lie roughly in the middle of the range from pale to dark, whereas intermediate adults are similar to darks and very different from pales.

If most intermediate chicks were heterozygous, becoming phenotypically darker as adults, dark adults would be either homozygous or

Table 5.1. *Chicks produced by matings in period 1951–58*

Mating type	Chicks produced			
	P	I	D	Total
P × P	29	0	0	29
P × I	36	42	3	81
P × D	16	39	2	57
I × I	16	37	8	61
I × D	9	125	43	177
D × D	0	12	15	27
All matings	106	255	71	432

In this table P represents pale, I intermediate and D dark.

heterozygous for the melanic allele. Alternatively, if most intermediate adults were heterozygous, while darks were homozygous, the relatively paler intermediate chicks would be either heterozygous or homozygous for the melanic allele. The homozygous intermediate chicks would presumably become dark adults. Unfortunately we cannot assume that the classification of phenotypes of either chicks or adults will correspond to genotypes.

In a first attempt to determine the genetics of the Arctic Skua, O'Donald & P.E. Davis (1959) observed chicks produced by different matings. They assumed that the chicks' phenotypes indicated their genotypes: the pale chicks were pale homozygotes; the intermediate chicks were heterozygotes; and the dark chicks were melanic homozygotes. In order to explain the appearance of intermediate, and presumably heterozygous chicks among the progeny of matings of dark × dark adults, they further assumed that a proportion λ of the dark adults were heterozygotes while the remaining $1 - \lambda$ were homozygotes. Table 5.1 shows the chicks produced by matings between the different adult phenotypes in the period 1951–58.

On O'Donald & Davis' assumptions, the three mating types, P × D, I × D and D × D involve both heterozygous and homozygous dark phenotypes. If d is the allele for pale and D the allele for melanism, the phenotypic mating type P × D consists of the two genotypic mating types

$$dd \times Dd \text{ and } dd \times DD$$

The mating $dd \times Dd$ has probability λ and should produce equal proportions of pale and intermediate chicks. The mating $dd \times DD$ has probability $1 - \lambda$ and only produces intermediate chicks. These matings should therefore produce chicks in the proportions

$$\tfrac{1}{2}\lambda \text{ (pale)}, \quad 1 - \tfrac{1}{2}\lambda \text{ (intermediate)}$$

Table 5.2 shows the expected frequencies of the progeny of all six phenotypic mating types. It is a simple matter to estimate λ by maximum likelihood using the data of table 5.1. O'Donald & Davis obtained the estimate

$$\hat{\lambda} = 0.454$$

This estimate is consistent for each of the matings $P \times D$, $I \times D$ and $D \times D$. A test of the heterogeneity of the estimates of λ to which these matings separately lead gives the value

$$\chi_1^2 = 1.4502; \quad P = 0.484$$

In this respect, the genetic model fits the data well. But the model cannot account for all the intermediate chicks that have been produced. Table 5.2 shows that matings $P \times I$, $I \times I$ and $I \times D$ should each produce intermediate and other chicks in the ratio $1:1$. From table 5.1 we see that the actual numbers of chicks produced by these matings are as follows

Mating	Intermediate chicks	Other chicks
$P \times I$	45	36
$I \times I$	37	24
$I \times D$	125	52
Total	207	112

The overall deviation from the $1:1$ ratio is of course very significant indeed:

$$\chi_1^2 = 28.2915$$

The deviation from the $1:1$ ratios also varies between the matings. A test of the homogeneity of the ratios in the 2×3 contingency table is just significant

$$\chi_2^2 = 6.1300, \quad P = 0.0467$$

If intermediate chicks are indeed heterozygotes, their excess over the expected proportions is very difficult to understand. It might imply that very strong selection takes place against both pales and darks in the period from hatching to fledging. After fledging, as we shall see in the following chapter (section 6.3), selection acts against heterozygotes, though only at low intensity. The deviation from the $1:1$ ratio at fledging would imply that pales and darks suffer a 46 per cent disadvantage compared to intermediates $(1 - 112/207 = 0.459)$.

Table 5.2. *Expected frequencies of pale, intermediate and dark progeny of each of the mating types*

Mating type	Expected frequencies of chicks		
	P	I	D
P × P	1	—	—
P × I	$\frac{1}{2}$	$\frac{1}{2}$	—
P × D	$\frac{1}{2}\lambda$	$1-\frac{1}{2}\lambda$	—
I × I	$\frac{1}{4}$	$\frac{1}{2}$	$\frac{1}{4}$
I × D	$\frac{1}{4}\lambda$	$\frac{1}{2}$	$\frac{1}{4}(2-\lambda)$
D × D	$\frac{1}{4}\lambda^2$	$\frac{1}{2}\lambda(2-\lambda)$	$\frac{1}{4}(2-\lambda)^2$

The expected frequencies in this table are calculated on O'Donald & Davis' assumption that a proportion λ of dark (D) adults are heterozygous (*Dd*), whereas pale (P), intermediate (I) and dark (D) chicks are always homozygous (*dd*), heterozygous (*Dd*) and homozygous (*DD*) respectively. The 5 dark birds produced by the P × I and P × D matings are impossible according to the genetic model. These birds have been assumed to be heterozygotes.

The heterogeneity in the deviations from the 1:1 ratio may arise from differences between pales and darks in their survival rates relative to intermediates. The greatest excess of intermediates is found in the I × D matings, which produced only nine pale chicks. The least excess is found in the P × D matings, which produced only two dark chicks. Suppose only the darks suffer reduced viability during the period after hatching. Two parameters must now be estimated: λ, and the viability of darks, v. At the maximum likelihood of this model, I estimate the values

$$\hat{\lambda} = 0.2874, \hat{v} = 0.4357$$

Thus, in this model, we see that the homozygous darks have less than 50 per cent of the viability of the other genotypes. After fitting the model to the numbers of progeny of the matings P × D, I × I, I × D and D × D, I find

$$\chi^2_4 = 13.1487, P = 0.0106$$

Significant heterogeneity still remains: the model does not fit the data. In an extension of the model, we may assume that both pale and dark birds suffer reduced viability. Then the additional mating, P × I, gives information about the viability of pale birds, ω. At maximum likelihood, I estimate

$$\hat{\lambda} = 0.3327, \hat{v} = 0.2294, \text{ and } \hat{\omega} = 0.8232$$

After fitting the model with these estimates,

$$\chi_4^2 = 13.0936, P = 0.0108$$

Both these values of χ^2 for the tests of goodness of fit of the models have four degrees of freedom, derived as follows. The matings $P \times I$ and $P \times D$ can each produce two types of progeny and each contribute one degree of freedom. The matings $I \times I$, $I \times D$ and $D \times D$ can produce all three types of progeny and each contributes two degrees of freedom. However, according to the model, the $D \times D$ mating should produce only a very small proportion of pale offspring, and in fact produced none: the pale and intermediate classes of offspring were therefore amalgamated leaving only two classes of progeny and only one degree of freedom for this mating. Before the fitting of any parameters, the five matings therefore contribute a total of seven degrees of freedom. After fitting all three parameters, four degrees of freedom are left. If only the two parameters λ and v are incorporated in the model, the $P \times I$ mating is no longer informative: the four remaining matings contribute six degrees of freedom, again leaving four after fitting the two parameters.

In their original paper, O'Donald & Davis (1959) failed to observe the significant heterogeneity in the ratios of intermediates to other phenotypes. They merely suggested that dark chicks were less viable than the others, but did not attempt to incorporate differential viability into the model. As we have now shown, however, the model cannot be salvaged even by assuming that all phenotypes differ arbitrarily in their viabilities. One obvious source of error may arise in the classification of the phenotypes of the chicks. They show a continuous range of phenotypic variation. According to the model, the $P \times I$ and $P \times D$ matings should produce no dark chicks, but some were classified as such (table 5.1). We must determine how far the classification of the chicks coincides with that of the adults. A total of 38 chicks, ringed from 1973 onwards, have been recaptured as breeding adults up to and including the 1979 breeding season. Eleven of these chicks were classified differently as adults. Twenty-six of the chicks were classified as intermediate or dark-intermediate. Four of these chicks were recaptured as-pale adults and one as a dark adult. Eight chicks classified as pale all returned as pale adults. Of the remaining four chicks which were classified as dark, one returned as an intermediate adult and two returned as dark-intermediate

Table 5.3. *Chicks produced by different matings in the period 1973–79*

Mating type	Chicks produced			
	P	I	D	Total
P × P	21	1	0	22
P × I	33	78	0	111
P × D	0	63	1	64
I × I	27	162	49	238
I × D	0	79	28	107
D × D	0	9	18	27
All matings	81	392	96	569

adults. These changes of classification may be expressed as follows

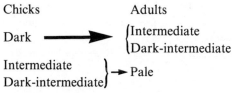

Chicks   Adults

Dark ⟶ { Intermediate
         Dark-intermediate

Intermediate }
Dark-intermediate } → Pale

The heavy arrow indicates the strong tendency of dark chicks to change their classification. The same criterion was used in classifying both chicks and adults: intermediates have belly feathers with dark tips and pale bases. Intermediate adults also have at least some pale or golden feathers on the sides of the neck. This is the character by which they can be recognized in the field. If dark chicks are mostly homozygous melanics, as O'Donald & Davis (1959) assumed, some intermediate adults, having been classed as dark chicks, must also be homozygous, having changed from being dark chicks into intermediate adults. This would contradict the premise of the model that intermediate adults are heterozygous while dark adults may be either homozygous or heterozygous. The original assumptions of O'Donald & Davis' model are clearly false. The model is refuted both by its poor fit to the data and by the changes of phenotype from chick to adult.

The classification of chicks is quite arbitrary, as we have seen, with no clear division between phenotypes. In the later period, 1973–79, when the Arctic Skuas of Fair Isle were studied in exhaustive detail, the phenotypes appear to have been classified somewhat differently. The data for this period are shown in table 5.3. All dark-intermediate birds have been included with intermediates. The numbers of intermediates exceed expectation by a greater proportion than in O'Donald & Davis's original data.

Table 5.4. *Chicks recaptured as adults produced by parents of known phenotype*

Mating type	Adult phenotype of chick				
	P	I	DI	D	Total
P × P	6	—	—	—	6
P × I	10	9	2	—	21
P × DI	1	2	—	—	3
P × D	—	8	2	—	10
I × I	7	11	2	2	22
I × DI	2	5	4	—	11
I × D	—	8	2	7	17
DI × DI	3	1	—	1	5
DI × D	—	2	1	4	7
D × D	—	—	—	3	3
All matings	29	46	13	17	105

In this table P represents pale, I intermediate, DI dark-intermediate and D dark.

Moreover, a far greater proportion of matings has been classed as I × I. We also know that several intermediate chicks have been recaptured as pale adults. Dark chicks have been recaptured as intermediate adults. One intermediate chick even became a dark adult. Since the classification of chicks as either intermediate or dark correlates only partly with their adult classification, while our division of the chicks into classes has also changed from 1959 to 1973, we must conclude that the classification of chicks is too uncertain, subjective and inconsistent for genetic analysis.

5.1.2 Chicks recaptured as breeding adults

Thirty-eight chicks ringed between 1973–75 were recaptured later as adults. Sixty-seven chicks ringed up to 1970 had earlier been recaptured as adults (O'Donald & J.W.F. Davis, 1975). Table 5.4 shows the data of the total of 105 chicks. The data of the 38 chicks have been added to the data of the 67, for the two sets of data show no heterogeneity. If dark-intermediates are included in the intermediate class, the data then fit the simple genetic model in which pales and darks are the two homozygotes and intermediates are the heterozygotes. Table 5.5 shows this classification of the data and the χ^2 tests of the fit of the model. The fit is reasonably good. In the matings

Table 5.5. *Adult phenotypes of offspring produced by different matings when dark-intermediates are classed with intermediates*

(i) Offspring of the matings

Mating type	Adult offspring produced			
	P	I	D	Total
P × P	6	—	—	6
P × I	11	13	—	24
P × D	—	10	—	10
I × I	12	23	3	38
I × D	—	13	11	24
D × D	—	—	3	3

(ii) Tests of goodness of fit to Mendelian ratios

Mating type	Value of χ^2	Dfs	Value of P
P × I	0.1667	1	0.683
I × I	5.9474	2	0.0511
I × D	0.1667	1	0.683

with segregating progeny, P × I, I × I and I × D, intermediates have still been produced in excess of the Mendelian expectations:

49 (intermediate): 37 (non-intermediate)

The expected ratio is 1:1 giving

$$\chi_1^2 = 1.6744, P = 0.196$$

This deficit of non-intermediate phenotypes arises largely from the small number of darks produced by the I × I matings. The ratio of phenotypes in the P × I and I × D is very close to the expected 1:1 ratio.

We know that a number of the intermediate chicks are homozygous for the pale allele. They become pale adults. Some intermediate adults are probably homozygous for the dark allele. We should expect that such homozygotes would tend to be the darker intermediates. But classifying dark-intermediates with darks as homozygous is incompatible with the genetic model, as table 5.6 shows. Then the P × I matings apparently produced some dark homozygotes, the P × D matings produced segregating progeny and the I × D and D × D matings produced homozygous pale progeny. These facts would be impossible if dark and dark-intermediate birds were homozygous. If some intermediates are indeed homozygous for

Table 5.6. *Adult phenotypes of offspring pro-duced by different matings when dark-interme-diates are classed with darks*

Mating type	Adult offspring produced			
	P	I	D	Total
P × P	6	—	—	6
P × I	10	9	2	21
P × D	1	10	2	13
I × I	7	11	4	22
I × D	2	13	13	28
D × D	3	3	9	15

the dark allele, they are not necessarily the dark-intermediates. Perhaps the expression of the heterozygotes and dark homozygotes may vary towards both lighter and darker pigmentation – some heterozygotes appearing dark. Considerable variation certainly occurs in the development of the adult phenotypes: as we have seen, dark chicks have become intermediate or dark-intermediate adults; intermediate chicks have become dark adults. The classification of table 5.5 has been explained by the simple genetic model, in which pale adults are homozygous for the pale allele, intermediate adults are mostly heterozygous, and dark adults mostly homozygous for the dark allele. In view of the uncertainties in classifying the melanic phenotypes, this is probably as much as we can say from these data about the genetics of the phenotypes of the Arctic Skua.

5.2 Frequencies of phenotypes in the Fair Isle population

In the period 1973–79 on Fair Isle, a total of 792 different birds were seen. These include all birds breeding in 1973 and all new birds arriving to breed in later years. Almost all birds were eventually ringed, most in their first breeding year. A pair with an unringed bird in one year was considered to be the same pair in the next year if it bred on the same territory and the unringed bird was of the same phenotype. Unringed birds are very unlikely to have been counted twice, but some new unringed birds may not have been counted. The 792 birds were of the following phenotypes:

pales	intermediates	darks
159	458	175

Evidently the intermediates greatly outnumber the others. If the interme-

diates are heterozygous, the gene frequencies can be estimated simply by
the following relative proportions

$$\hat{p} = \frac{159 + 458/2}{792}$$
$$= 0.4899 \text{ (pale)}$$
$$\hat{q} = 0.5101 \text{ (melanic)}$$

But, if the frequencies of pale, intermediate and dark follow the Hardy-
Weinberg Law, p^2, $2pq$, q^2, the numbers should be

pales	intermediates	darks
190.081	395.838	206.081

giving

$$\chi_1^2 = 19.5317, \ P = 9.89 \times 10^{-6}$$

The Hardy-Weinberg Law does not hold. It is based on the assumption that
individuals mate at random with no selection acting upon them. We know
this is not true: melanics mate assortatively (chapter 10); intermediates are
at a slight selective disadvantage after fledging (chapter 6, section 6.3, see
table 6.9). Neither of these effects, however, would produce the observed
excess of intermediates over their expected Hardy-Weinberg frequency.

If, as I have already suggested, some intermediate adults are homozygous for the dark allele, the gene frequency cannot be determined directly
from the numbers of the phenotypes. Assuming the Hardy-Weinberg Law
holds approximately, the hypothetical frequency of the homozygous
recessive can be equated to its actual proportion in the population. We
know that pales are homozygous, so we can equate

$$p^2 = 159/792$$

to obtain

$$\hat{p} = 0.4481$$

This is smaller than the estimate based on the assumption that all
intermediates are heterozygous. The frequency of heterozygotes should be

$$2\,\hat{p}\,\hat{q} = 0.4946$$

As intermediates their actual frequency is greater than this prediction:

$$458/792 = 0.5783$$

If some intermediates are homozygous for the dark allele, it follows that the

I × I matings must consist of three different genotypic types:

$Dd \times Dd$

$Dd \times DD$

$DD \times DD$

About 14 per cent of intermediates may actually be homozygous, DD, since

$(0.5783 - 0.4946)/0.5783 = 0.1447$

Allowing for this proportion of homozygous intermediates, the offspring of the I × I matings should be produced in the genotypic ratios

$dd:Dd:DD::0.1829:0.4895:0.3276$

Hence, the phenotypic ratios should be

$P:I:D::0.1829:0.5369:0.2802$

More darks than pales should appear in the progeny. But many fewer darks actually appeared (table 5.5). It is not strictly valid to calculate a value of χ^2 from these hypothetical phenotypic ratios since an estimate – the proportion of homozygous intermediates – was used in their calculation. But this estimate is completely independent of the numbers of progeny produced by the I × I matings; and it has been derived from quite a large sample. As a rough indicator of statistical significance, I calculate that

$\chi^2 \simeq 9.5, P \simeq 0\cdot009$

The data appear to refute the hypothesis fairly decisively. At every next step, we seem to meet a contradiction of our previous hypothesis!

Perhaps we can only conclude, as at the end of the last section, that pale adults are homozygous for the pale allele, intermediates are mostly heterozygous, and dark adults are mostly homozygous for the melanic allele.

5.3 Heritability of breeding dates

Breeding date is a 'quantitative' character. In principle, it could be any instant in the continuous temporal sequence of the breeding season. Simple Mendelian factors are each separately determined by unique sequences of the nucleotides in the genetic code: they are discrete, not continuous. Alas! As we have seen, the phenotypic expression of simple Mendelian differences can vary so much as to become a continuous character – like the phenotypes of the Arctic Skua chicks. Breeding date is the product of environmental factors, particularly daylength, interacting with the physiology and behaviour of the animal. Such characters are usually assumed to depend on the combined effects of many different genes.

The genes contribute some of the variation, the environmental factors the rest. Since the effects of individual genes cannot be discerned, the study of inheritance becomes an analysis of the components of variation: how much do the genes contribute to the variation and how much does the environment? There is no sense in which a particular individual's characteristic value – his breeding date for example – can be apportioned to heredity and environment, for both were necessary to his development. The apportionment is of the variation between individuals, not of an individual's particular value. This leads to an analysis of variance.

Variation is measured by a statistic called the variance, or its square root, called the standard deviation:

$$\text{var}(x) = s^2 = \Sigma(x-\bar{x})^2/(n-1)$$

where $\bar{x}$ is the mean of observations in a sample of n. An alternative formula is

$$s^2 = [\Sigma x^2 - (\Sigma x)^2/n]/(n-1)$$

This latter formula is much simpler to use for calculation, since it does not involve finding the squares of the deviations, $x - \bar{x}$. These formulae estimate the population variance σ^2 – a parameter of the normal and other theoretical distributions. The estimate is not simply the 'mean squared deviation', for the sum of the squared deviations is divided by the number of degrees of freedom, $n-1$, not n. As an estimate of the population variance, the mean squared deviation

$$S^2 = \Sigma(x-\bar{x})^2/n$$

is biassed. This means that if a large number of samples of size n were taken, the average value of S^2 would show a small, consistent deviation, or bias, from σ^2. Dividing by $n-1$, not n, eliminates the bias: the average value of s^2 would equal σ^2. By taking the square root of the variance, we find the standard deviation, s. But this is not a useful measure of variation: it is not additive; standard deviations corresponding to different components of variation cannot be added together to give the overall standard variation. Components of the variance sum to the total variance. Thus we can obtain an 'analysis of variance', but not an 'analysis of standard deviation'.

The variance of a character like breeding date may be comprised of many separate components:

(i) the effect caused by different individuals having different genes, known as the 'additive genetic' effect;
(ii) the effects of dominance and gene interactions like epistasis;
(iii) the interaction of genes with the environment, some genes

producing one effect in one environment and another effect in another environment;

(iv) various environmental effects acting either generally or only within families.

All these factors may contribute to the variance in the phenotypic expression of the character. We may write, for example,

$$V_p = V_A + V_D + V_I + V_{GE} + V_E$$

The symbols stand for the different components of variance:

V_p for the overall phenotypic variance
V_A for the additive genetic variance
V_D for the dominance variance
V_I for the variance of genetic interaction
V_{GE} for the variance of genotype-environment interaction
V_E for the environmental variance.

A detailed breeding programme is necessary to determine these components of variance. To an animal breeder or population geneticist, V_A is the most important component. It determines how selection will change a population. Provided there is some additive genetic variance ($V_A \neq 0$), selection for high or low values of the character will select genes for high or low values. In the next generation, the mean value of the character will have increased in a line selected for high values, or decreased in a line selected for low values: a response to selection will have occurred. But no such response can occur if $V_A = 0$. So, if a breeder wishes to select for an increased yield, for example, yield must be a character showing additive genetic variation. Otherwise he will get nowhere. In general, evolution can occur only if $V_A \neq 0$. Those Marxist sociologists who assert that human intelligence has no inherited component must believe in special creation, at least for this human trait.

The concept of heritability, given the symbol h^2, can now be defined. It is that fraction of the total variance which is determined by additive genetic effects

$$h^2 = V_A/V_p$$

It is shown in textbooks on quantitative genetics (for example, Falconer, 1960, 1980) that h^2 can be estimated from the average effects of parental values on the values of their offspring. Statistically, these average effects are the regression coefficients of offspring values on parental values. We can calculate the regression of offspring mean on the mean of both parents or the regression of offspring mean on one parent only. Making a number of

Table 5.7. *Breeding dates of parents and offspring*

Chick no.	Breeding dates		Chick no.	Breeding - dates	
	Parent	Chick		Parent	Chick
730	32	28	667	14	24
669	23	33	651	13	16
566	24	32	804	33	50
—		23	819	43	38
528	32	37	735	16	30
513	17	23	702	14	34
510	20	30	699	13	41
506	43	27	568	32	37
981	28	33	813	38	41
754	17	23	811	32	38
683	16	32	796	28	37
—		19			
682	13	23			

In this table, Chick no. is the last three digits of the chick's ring no. The breeding dates are hatching dates in days after 1 June. In two cases, as shown, the chick bred with a new mate in two successive occasions for which breeding dates could be ascertained. Eleven chicks were females and 12 males. Each sex showed a very similar regression on the parental values ($b_{o,p}=0.322$ for female chicks, $b_{o,p}=0.435$ for male chicks). This is surprising: females most probably determine breeding date and may be expected to show a higher regression. But sample sizes are very small: differences in the regressions of males and females could hardly be detected.

assumptions (see Falconer, 1980, for details), we can then estimate

$$h^2 = b_{\bar{o}, \bar{p}} \text{ (regression of offspring mean on parental mean)}$$
$$h^2 = 2b_{\bar{o}, p} \text{ (regression of offspring mean on one parent)}$$

Normally, if offspring means are regressed on the values of the parent, male offspring are regressed on the value of the male parent and female offspring on the value of the female parent. This allows for any difference in the character between the sexes.

Breeding dates of both parents and offspring were obtained for most of the 38 chicks that fledged since 1973 and returned to Fair Isle to breed. Breeding date is a characteristic of pairs, both in parents and offspring. The heritability is then

$$h^2 = 2\,b_{o,p}$$

Table 5.7 gives the data of the hatching dates used to measure the breeding dates of the parents and offspring. Breeding dates can be observed for several years. But after the first year, the dates of a particular pair become earlier and then remain almost identical from one year to the next. In calculating the regression of offspring on parents I have only used the first date when the pair bred together. The table gives values of 24 dates for parents and offspring. The regression coefficient is estimated by

$$b_{o,p} = 0.4267$$

Figure 5.1 Breeding dates of parents and offspring showing regression of offspring value on parental value. Breeding date is measured by hatching date in days from 1 June. Squares represent males, circles females.

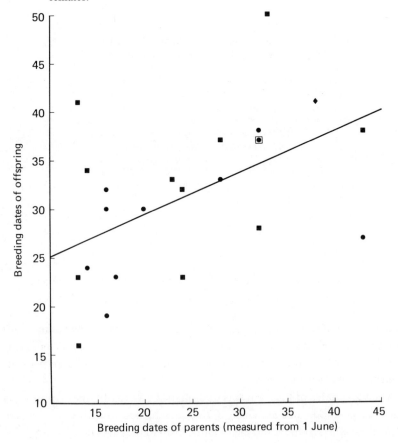

and hence approximately the heritability

$$h^2 = 0.85$$

Figure 5.1 shows the clear relationship between parents and offspring in their breeding dates. The heritability is high and very significant as the following analysis of variance of the regression shows:

Source of variation	Sum of squares	Degrees of freedom	Mean square	Value of F	Value of P
Regression	392.551	1	392.55	7.898	0.0102
Residual	1093.407	22	49.700	—	—

A high heritability such as $h^2 = 0.85$ could produce a very rapid response to strong selection. And we know selection does act strongly on breeding dates. Early breeding pairs are much more successful than late pairs. Breeding date is closely correlated with fitness as table 8.7 (chapter 8, section 8.5) shows. Are breeding dates therefore getting generally earlier? They could not do so indefinitely, for, sooner or later, Arctic Skuas would be arriving before the other seabirds and therefore before food was available. The early breeders would then suffer a heavy penalty in their turn: when early and later breeding are both equally disadvantageous, stabilizing selection will act to maintain an optimum mean breeding date. There seems to be no evidence that the Arctic Skua has been breeding earlier in recent years. According to Witherby *et al.* (1941)

'Summer residents return to nesting grounds end April (in some years from third week) to fourth week May. Nesting places begin to be deserted mid-Aug. and most abandoned by first week Sept'.

Arctic Skuas on Fair Isle still keep to this schedule of arrivals and departures (chapter 4, section 4.1). The greater reproductive advantage of early breeding has not been translated into evolutionary change, in spite of the high heritability of breeding date. Why not? We can only speculate on the answer. It would be highly disadvantageous for Arctic Skuas to return to their breeding grounds before the Kittiwakes and Puffins which are the main targets of the Arctic Skuas' piracy (chapter 3, section 3.1). Consider the effects on fitness of Arctic Skuas returning to breed over a wide range of different dates from very early to very late in the season. Before the other seabirds had returned, their fitness would be very low or zero. While the other seabirds were returning, fitness would rise sharply. It would soon reach an optimum and then gradually fall away through the later part of the season, for late chicks have a poor chance of survival. The fitness function of breeding date must be extremely asymmetrical, truncated at the

Table 5.8. *Clutch size of parents and offspring*

Offspring's clutch size	Parent's clutch size			
	0	1	2	Total
0	—	—	1	1
1	—	1	12	13
2	—	1	25	26
3	—	—	1	1
Total	—	2	39	41

beginning, extended into a long tail at the end. Very strong selection would eliminate any birds attempting to breed before fitness had started to rise. Selection would have no net effect of pushing the Arctic Skuas into earlier and earlier breeding.

5.4 Heritability of clutch size and fledging success

Clutch size and fledging success are components of fitness. They determine the values of b_x – the age-specific reproductive rates – hence the intrinsic rates of increase and hence the fitnesses (see appendix A). Quantitative geneticists have argued that fitness can have only low heritability (Falconer, 1960). Since genes that directly determine the components of fitness are rapidly selected, they will usually have reached equilibrium or near equilibrium. The additive genetic variance should therefore be small and the heritability low. This argument is sometimes used to deny evolution. For example, in the theory of sexual selection, females who mate preferentially with fitter males produce fitter offspring. This selects the mating preference (see chapter 9, section 9.1 for a detailed account of this theory). Some critics have argued that since heritability in fitness must be low, female preference cannot evolve. The same argument could be employed to deny the evolution of any character. Yet, to take a simple example, we know that insects can produce mutations giving resistance to pesticides. These mutations are very rapidly selected in areas where pesticides are used: they provide the genetic variance in fitness on which the selection acts. When the population has become fully resistant, the genetic variance in fitness will have largely disappeared. The heritability of fitness will be high only while rapid evolution is taking place. But this does not imply that selection will have no genetic variation to act upon at other times.

In theory, an organism's reproductive rate will be selected to maximize the net contribution of offspring to the next generation. David Lack found that some species of birds produced an optimum number of eggs (Lack, 1954). If they laid more eggs than the optimum, they actually raised fewer chicks because they could not feed the extra number successfully. They also, of course, raised fewer chicks if they laid fewer than the optimum number. Lack assumed, as many ecologists do, that the optimal expression of a character implies that it must have evolved by natural selection. This selection must have had genetic variance to act upon.

I ascertained the clutch size and fledging success both of the 38 chicks that had been recaptured as breeding adults and of their parents. I only counted such data when the birds had formed new pairs. Some birds formed new pairs in successive years, giving more than just one observation. The data of clutch size are shown in table 5.8 and fledging success in table 5.9. Clutches usually contain two eggs or one. There is so little variation between parents and offspring that the correlation is inevitably small and not significant:

$$r = 0.063$$

To detect a correlation between parents and offspring, data would have to be obtained from a large number of chicks that returned to breed.

Parents and offspring show more variation in fledging success than in clutch size (table 5.9). But the correlation and regression coefficients

$$r = 0.165, b = 0.265$$

are not statistically significant. The 2×2 table of the data

		Chicks fledged by parents	
		0 or 1	2
Chicks fledged by offspring	0 or 1	5	8
	2	4	19

gives the value of χ^2 with Yates correction

$$\chi_1^2 = 1.0033, P = 0.317$$

This result is just what we should expect by random sampling. It gives no evidence for heritability.

If clutch size and fledging success do show heritable variation, it will be almost impossible to detect it given the numbers of observations that could be made in practice. The variation in clutch size, heritable or not, is always

Table 5.9. *Fledging success of parents and offspring*

Offspring's fledging success	Parent's fledging success			
	0	1	2	Total
0	—	5	8	13
1	1	1	10	12
2	—	2	9	11
Total	1	8	27	36

very small, since most birds lay just two eggs. A pair should easily be capable of rearing more than two chicks, for a single bird whose mate has died can usually manage to rear two chicks by itself. Yet clutches of three eggs have never been successful. Apparently a bird with three eggs can only incubate two of them at a time. One is left unincubated. We have deduced this by observing that when a bird with three eggs is put off its nest, one of its eggs is always found to be cold, presumably not having been incubated. If different eggs in a clutch of three became chilled in turn, all would eventually die. Clutches of four eggs have been reported (Witherby *et al.*, 1941), but never observed on Fair Isle. Once, we removed a newly laid third egg from a clutch of three and placed it in a nest containing only one egg, hoping it would hatch after being continuously incubated. But it was eventually broken and lost. No chicks have ever been produced from clutches of three eggs.

A clutch of two eggs has the overwhelming advantage: a bird would never gain an advantage by consistently producing only one egg; clutches of three or zero represent complete infertility. Clutches of one are usually repeats following the loss of a first clutch, or produced by new birds late in the season. The heritability of clutch size should have been reduced to a low value or zero. But some heritable variation in fledging success should still exist. Fledging success varies more than clutch size: it varies with breeding date; it depends on the abilities of the parents – their abilities to incubate their eggs and protect and feed their chicks. Some parental activities conflict with others: chicks are not being protected while their parents are away collecting food. Selection is unlikely to have eliminated all inherited components of the variation in the different parental behaviours. Since fledging success declines sharply late in the breeding season, the high heritability in breeding date should produce some heritability in fledging success. As we have seen, however, my own data are insufficient to detect any significant heritability at the level of the observed regression of offspring on parents. At this level ($b_{o,p} = 0.265$), I should have had to collect twice as much data in order to have obtained a statistically significant result.

6

Demography and selection

The demographer collects data of birth and death rates so that he can predict population trends and changes. He needs to know the probabilities of surviving to different ages and the mean number of offspring born to individuals in each age class. Such data are very difficult to obtain except from human populations. Ideally, the probabilities of survival should be estimated by the proportions of survivors of a cohort of individuals who have been followed from birth until all have died.

Human populations consist of individuals of all ages. Reproduction is continuous. Many organisms reproduce annually at a particular time of the year. A population can then be divided into separate age classes. When the chicks have hatched in a population of birds, for example, individuals may be 0, 1, 2, ... or x years old, having either just hatched, or hatched last year, or two years ago or, in general, x years ago. Birds are easily counted or caught for ringing in the breeding season. This is convenient because they are thus counted when almost exactly 0, 1, 2, ... or x years old.

The demography of a population is usually described in terms of a standard set of symbols. The symbols $l_0, l_1, l_2, \ldots l_x$ denote the probabilities of surviving for 0, 1, 2, ... or x years. Obviously, $l_0 = 1$, since the new-born have had no time to die. At each of these ages, individuals produce $b_0, b_1, b_2, \ldots, b_x$ offspring on average. Obviously, $b_0 = 0$, since the new-born do not immediately reproduce. A bird aged $x - 1$ years survives to age x with probability l_x/l_{x-1}. It then produces b_x chicks. Of course, it does not produce them instantaneously, at the precise moment when it becomes x years old. The breeding season may last two or three months. The bird may die during this period whilst producing its b_x chicks. Survival and reproduction are not completely separate events. Fortunately, they can be treated separately. Only a small proportion of Arctic Skuas die during their breeding season. Both birds of a pair are most unlikely to die in the same season. If one parent dies, the other can still incubate the eggs and rear the chicks. Single Arctic Skuas are usually successful parents and fledge their

120

chicks. The dead parent produces its b_x chicks posthumously. The small proportion of deaths in the breeding season can therefore be regarded as part of the mortality that occurs in the interval before the next breeding season. We can divide the life of an Arctic Skua into a succession of annual cycles within which it survives from one breeding season to the next with probability l_x/l_{x-1} and then reproduces at rate b_x. As with all sexually reproducing organisms, b_x is half the average number of offspring produced by a pair.

On Fair Isle, the exact numbers of breeding pairs of Arctic Skuas have been ascertained for each of the years 1948–62 and 1973–79. From 1948 to 1962, the population increased from 15 to 71 pairs. As we saw in section 2.2 of chapter 2, the equation

$$N_t = 16.62 \exp (0.1092\ t)$$

fits the observed numbers of pairs. In this equation t represents the time in years, starting in 1948 at $t=1$ and ending in 1962 at $t=15$. The general equation

$$N_t = N_0 e^{rt}$$

describes an exponential increase or decrease in population numbers. The rate of increase

$$\frac{dN_t}{dt} = rN_0 e^{rt} = rN_t$$

shows that

$$r = \frac{1}{N_t} \left(\frac{dN_t}{dt} \right)$$

$$= \frac{d \ln (N_t)}{dt}$$

The rate of change of the logarithm of population size is thus equal to the constant r. This is called 'the intrinsic rate of population increase'. It measures the instantaneous rate of increase occurring in any small interval of time. The quantity

$$\lambda = e^r$$

measures the increase in numbers that takes place during one complete unit of time – for example, the increase from one breeding season to the next. Thus if

$$r = 0.1092$$

then

$$\lambda = 1.115$$

The percentage increase in one year is therefore

$$100 (\lambda - 1) = 11.5\%$$

We can also measure the population increase or decrease that occurs in one generation. If T is the duration of a generation in years, then

$$R_0 = e^{rT}$$

R_0 is the average number of offspring that an individual contributes to the next generation. It is determined by the values of l_x and b_x. Consider a chick that has recently hatched. How many chicks will it produce in its turn? Its chance of surviving to age x is l_x. If it survives it produces b_x chicks. In its xth year of life, an average chick will therefore produce $l_x b_x$ chicks of its own. Its total production of chicks is the sum of the numbers it produces each year. This sum is the average contribution that one chick will make to the chicks of the next generation. If the number of chicks is increasing at the same rate as the population as a whole (the age structure thus remaining the same), we can then equate

$$R_0 = e^{rT} = \sum_x l_x b_x$$

These equations are used to estimate the generation time, since

$$T = (\ln R_0)/r$$

As shown in appendix A, the value of r is also calculated from l_x and b_x.

Selection occurs when individuals differ in their survival or reproductive rates. Since all individuals differ, everyone may have their own characteristic values of l_x and b_x. But these can never be individually determined. We can determine l_x and b_x only for particular groups of individuals: l_x and b_x are parameters; they are characteristics of individuals; but samples must be taken from the population to estimate them. For example, we might estimate l_x and b_x for a group of individuals who possess a particular genotype. If different genotypes have different values of l_x and b_x, they will also differ in their intrinsic rates of increase. One genotype will increase in frequency relative to another. Selection will give rise to evolution. In population genetics, a generation is the unit of evolutionary time. Selection is measured by the relative numbers of offspring that different genotypes contribute to the next generation. Relative fitness is defined as the ratio of the contributions of one genotype to another. Given the fitnesses of the genotypes, a population genetic model can be used to predict the genotypic frequencies in the next generation.

Suppose that l_{1x} and b_{1x}, l_{2x} and b_{2x} are the survival and reproductive rates of two different genotypes, which therefore produce average numbers of offspring

$$R_0^{(1)} = \sum_x l_{1x} b_{1x}$$

$$R_0^{(2)} = \sum_x l_{2x} b_{2x}$$

It might be thought that the relative fitness would be the ratio of these numbers. But this would not be correct. The two genotypes probably differ in their expectation of life and generation time. We need to find the relative numbers of offspring produced in the course of one generation of the whole population. The mean generation time of the population is given by

$$T = (\ln R_0)/r$$

where R_0 and r are the overall population values. Then if r_1 and r_2 are the intrinsic rates of increase of each genotype, the relative fitness of one genotype compared to the other is defined by the equation

$$w_1 = e^{r_1 T}/e^{r_2 T}$$
$$= e^{(r_1 - r_2)T}$$

By convention, fitness is measured relative to the fittest genotype. Thus, if $r_2 > r_1$, genotype 2 is the fitter, and we should have

$$w_2 = 1$$
$$w_1 < 1$$

Genotype 1 is less fit by the amount

$$s = 1 - w_1$$

This is called the 'selective coefficient'. It measures a genotype's relative disadvantage compared to the fittest genotype.

6.1 Survival rates of the phenotypes

Our aim in analysing data on survival rates of Arctic Skuas is to obtain estimates of l_x for each phenotype. They will be needed for the calculation of intrinsic rates of increase and fitnesses. Unfortunately, it would be impractical to attempt to follow the fortunes of a cohort of newly fledged chicks. The numbers of chicks produced in any one colony of Arctic Skuas would be too small. They would not necessarily return to their own natal colony but might move to other colonies in Shetland. They could not be caught and identified before their first breeding season. Even an army of helpers could hardly ring every Arctic Skua chick in the whole of Shetland. They could not possibly catch and identify every ringed adult to see if it was one of the cohort of chicks. Even if they could, a certain proportion of chicks may have emigrated from Shetland. No data could be obtained on the survival of chicks in the years before they had come back to breed. These insuperable difficulties would be encountered in attempts to observe the survival rates of most birds. Less direct methods must be used.

One method is to observe the age distribution of the population and assume that this reflects the probabilities of survival. For example, Bulmer & Perrins (1973) observed the age distributions of breeding male and female Great Tits in a population near Oxford. They found that a geometric distribution fitted the numbers of birds aged 1, 2 or more years old. Such a distribution would arise if death rates were constant and independent of age. If so, the probability of surviving from one year to the next would also be constant. We should have

$$l_x/l_{x-1} = 1-d$$

for all x, where d is the mortality in any one year. Since $l_0 = 1$, this gives

$$l_1 = 1-d$$
$$l_2 = (1-d)^2$$

and in general

$$l_x = (1-d)^x$$

These are probabilities of survival in a cohort of individuals. Bulmer & Perrins observed the numbers of individuals aged one year or older. To convert the probabilities of survival into proportions of individuals of different ages, we must divide each value of l_x by their sum

$$\sum_{x=1}^{\infty} l_x = (1-d)/d$$

The frequency distribution of individuals aged 1, 2, ... or x years will therefore be

$$f(x) = d(1-d)^{x-1}$$

This represents the geometric distribution which fits the data. Besides the assumption that

$$l_x = (1-d)^x$$

several other assumptions are implicit in this derivation. The constant mortality, d, must have occurred over a period at least equal to the age of the oldest birds. The same number of offspring must have been produced each year. If the population size and number of offspring increased, for example, more younger birds would be found in the population than the survival rate would predict. Their presence would produce too high an estimate of d, for it would seem that more older birds had died. If immigrants have entered the population, they must show the same age distribution as the population itself. Otherwise, the age distribution would not reflect the values of l_x.

Since the Arctic Skuas on Fair Isle increased exponentially between 1948 and 1962, the small number of very old birds must all have been produced when the colony was small. Most of the young birds were produced after it had expanded. The age distribution is therefore heavily biassed towards the younger birds. In an early paper on the demography of the population (O'Donald & Davis, 1976), I attempted to compensate for this by inversely weighting the numbers of birds of each age according to the size of the population from which they might have come. I ignored the problem of migration. But we now know that at least 45% of the new breeding birds were immigrants in 1978 and 1979. Unless the number of immigrants increased at the same rate as the colony, the correction for the bias in the age distribution would be invalid. Unfortunately, I cannot estimate the immigration rates in the earlier period, 1948–62, because the soft aluminium rings we used wore rapidly and dropped off the birds' legs after a few years. Many chicks lost their rings and would not have been identifiable as having hatched on Fair Isle. Unless we know how migration has varied, we cannot correct for its effects. If other colonies had increased in size at the same rate as the Fair Isle colony, the migration would probably have increased at the same rate too. The correction for bias would be valid. But in the period 1948–62, other colonies do not seem to have shown the same increase as on Fair Isle. Numbers on Foula, Noss and Hermaness increased only slightly or not at all (see tables 2.1, 2.2 and 2.3 given in section 2.2, chapter 2). The assumption implicit in the correction for bias is certainly unjustified, even though the actual magnitude of the correction to the data seems to have little effect on the estimates of the relative fitnesses and selective coefficients of the phenotypes.

We can justifiably assume that Arctic Skuas have a constant annual mortality like Great Tits. The survivorship of many species can be described by the general equation

$$l_x = (1-c)(1-d)^{x-1}$$

Thus

$$l_1 = 1-c$$
$$l_2 = (1-c)(1-d)$$
$$l_3 = (1-c)(1-d)^2$$

and so on. This allows for the fact that the mortality in the first year of life is usually greater than the constant mortality of the later years. In the analysis of the Great Tit data, the first year mortality was irrelevant since newly fledged birds were not included. Even if the first year mortality, c, was greater than the annual mortality in later years, d, we still obtain the simple geometric distribution for birds aged 1, 2, ... or x years:

$$f(x) = l_x / \sum_{x=1}^{\infty} l_x \quad (x = 1, 2, \ldots)$$

$$\sum_{x=1}^{\infty} l_x = (1-c)/d$$

Therefore

$$f(x) = d(1-d)^{x-1}$$

as previously obtained.

The value of d can easily be estimated, both for the population as a whole, and for each phenotype, dark, intermediate and pale: since the adults are ringed, we simply find the proportions that survive from one breeding season to the next. But birds cannot be caught in the years before they breed. The probabilities of surviving in the years after fledging and before breeding can only be determined by an indirect method. As shown in appendix A, about 34.6 per cent of the chicks survive to breed. This estimate is based on the proportion of immigrants into the population, hence the proportion of ringed chicks that would have returned to breed on Fair Isle, and hence the proportion that actually did so. If a bird usually breeds at four years old, then we have the equation

$$(1-c)(1-d)^3 = 0.346$$

Thus if

$$d = 0.2$$

then

$$c = 0.32$$

as shown in appendix A. This is a reasonable figure for the first-year mortality of a seabird. More accurately, we should use the actual distribution of age at maturity and the estimated annual mortality for 1948–62

$$d = 0.1994$$

giving

$$c = 0.277$$

The estimate of the first-year mortality depends critically on the assumption that we ringed all chicks fledged on Fair Isle. The estimate of c is reduced if we missed a proportion of the chicks. The estimate of c is very rough, but its exact magnitude has hardly any effect on the estimates of the selective coefficients.

I have just hinted at another basic assumption on which my demographic calculations depend. The proportion of chicks that survive to maturity is not a proportion of the chicks that hatch but a proportion of the chicks that fledge. In my calculations, the values of b_x are numbers of chicks fledged, not numbers hatched. Chicks are usually ringed in the week before fledging. They cannot be ringed immediately they hatch. The subsequent survival is measured in terms of the numbers of fledglings that survive. This has the effect of including the mortality of chicks before fledging with failure to hatch. In all the following demographic calculations, reproductive rate is fledging rate. The annual cycle is thus assumed to run from the end of one breeding season, when chicks were fledged, to the end of the next. Breeding birds dying in the breeding season are still included in the mortality of the following year, for, as explained, their chicks will normally be reared and fledged by the surviving parent.

O'Donald, Wedd & Davis (1974) published data of the annual survival rates of adult Arctic Skuas in the period 1948–62. Table 6.1 reproduces their table of the original data. The analysis of χ^2 in the table shows no significant differences in the survival rates of the three phenotypes. Selection does not appear to depend on differences in survival rates. This conclusion is reinforced by the survival rates in the period 1973–78: table 6.2 shows the numbers of birds that failed to return to the colony in the following breeding season and the numbers that were found dead in the current breeding season. The overall rates from 1973 to 1978 are very similar for each phenotype:

	Pale	Intermediate	Dark
Dead	57	151	46
Survived	214	682	217
Percentage mortality	21.0	18.1	17.5

$\chi_2^2 = 1.3973, P = 0.497$

6.1.1 Effects of shooting on survival in the Fair Isle population

The average mortality increased rapidly after 1975 as the following table, condensed from table 6.2, shows:

Table 6.1. *Survival of phenotypes from one breeding season to the next in the period 1948–62*

(i) Proportions of survivors

Phenotype	Females		Males		Both sexes	
	Proportion surviving	No. in sample	Proportion surviving	No. in sample	Proportion surviving	No. in sample
Dark	0.7391	46	0.8585	106	0.8224	152
Intermediate	0.7857	224	0.7978	183	0.7912	407
Pale	0.8202	89	0.7755	49	0.8043	138
All phenotypes	0.7883	359	0.8136	338	0.8006	697

(ii) Analysis of χ^2

Component of variation	Value of χ^2	Dfs	Value of P
Differences between the sexes in phenotypic proportions	38.8112	2	3.60×10^{-9}
Differences between phenotypes in survival rate	0.6907	2	0.708
Differences between sexes in survival rate	0.6984	1	0.403
Interaction of sex and phenotype on survival rate	3.2024	2	0.202
Total variation in $3 \times 2 \times 2$ table	43.4027	7	2.78×10^{-7}

Table 6.2. *Survival of phenotypes from one breeding season to the next in the period 1973–78*

Year	Birds that did not return			Birds found dead			Total number in colony		
	P	I	D	P	I	D	P	I	D
1973	3	16	5	2	1	—	35	111	39
1974	3	7	4	—	2	—	46	135	40
1975	5	21	2	—	3	1	51	157	46
1976	14	17	9	2	5	—	53	144	42
1977	14	30	13	3	8	2	53	159	57
1978	9	36	9	2	5	1	33	127	39
All years	48	127	42	9	24	4	271	833	263

In this table, the birds that did not return are those that were present in the year indicated, but not present in the subsequent year or years. The birds found dead are those that died in the breeding season of the year indicated. The final year of the project, 1979, is not included, because the subsequent survival of that year's birds is not known. Only identified ringed birds are included in this table.

Year	Number of birds			Percentage mortality	
	Survived	Dead	Total		
1973	158	27	185	14.6	
1974	205	16	221	7.2	11.36
1975	222	32	254	12.6	
1976	192	47	239	19.7	
1977	199	70	269	26.0	25.32
1978	137	62	199	31.2	
All years	1113	254	1367	18.6	

The following analysis of χ^2 of these data

Component of variation	Value of χ^2	Dfs	Value of P
Between years 1973–75	4.0168	2	0.134
Between years 1976–78	9.6191	2	0.00815
Between periods 1973–75 and 1976–78	43.9378	1	1.2×10^{-9}

shows that in the period 1973–75, mortality did not vary significantly from a mean of 11.36 per cent. After 1975 it increased sharply and continued to increase significantly. Sadly, the cause of this increase is known. Until 1975, John Davis worked full-time as my assistant on the Arctic Skua project. He spent the whole of the breeding season on Fair Isle. The Arctic Skuas were under his continual surveillance. From 1976 onwards, I was assisted part-time by John and others. The Arctic Skua colony was less closely watched. Many of the birds found dead had apparently been shot. We had usually left Fair Isle once the chicks had been ringed. The skuas could then have been shot unobserved. This would explain the increasing proportion of birds that failed to return to the colony next year. I ended the project in 1979, partly because I was disillusioned by the prospects for further demographic studies.

The present level of shooting will exterminate the Arctic Skuas of Fair Isle. We can make two estimates of the annual adult mortality before 1976. For the period 1948–62

$$d = 0.1994$$

As we have seen, this leads to an estimate of the first-year mortality

$$c = 0.277$$

If, by shooting breeding birds, the annual mortality were raised to

$$d = 0.312$$

as in 1979, then the reproductive rates and distributions of age at maturity (appendix A) would produce an intrinsic rate of increase

$$r = -0.127$$

The mortality for 1973–75

$$d = 0.1136$$

gives the estimate

$$c = 0.482$$

Then, if d is raised to 0.312,

$$r = -0.170$$

At this high negative rate, the Arctic Skua colony will be reduced from 114 pairs in 1979 to 21 pairs in 1989 and 4 pairs in 1999. Even at the lower negative rate, $r = -0.127$, the colony would be reduced to 32 pairs in 1989 and nine pairs in 1999. Since the authorities have not been influenced by argument in this matter, it will be interesting to see whether the National Trust for Scotland and Fair Isle Bird Observatory Trust will condone the possible extermination of one of Britain's rarest seabirds.

The variation in adult mortality gives rise to a problem: that of choosing an appropriate value to calculate R_0 and r. We have found three very significantly different mortality rates:

Period	Percentage mortality
1948–62	19.94
1973–75	11.36
1976–78	25.32
1948–78	19.04

Fortunately, in none of the data on mortality is there any evidence of differences in the mortality of the phenotypes. The deliberate shooting in the later years, 1976–78, was probably indiscriminate; if so, it may have obscured differences in phenotypic mortality rates. But there is no evidence of any variation in mortality in the preceeding period 1973–75:

	Pale	Intermediate	Dark
Dead	13	50	12
Survived	119	353	113
Percentage mortality	9.8	12.4	9.6

$$\chi_2^2 = 1.1224, P = 0.571$$

Table 6.3 shows the complete data on the mortality and survival of the phenotypes. The analysis of χ^2 given in the table reveals only one significant effect: the difference in mortality between the periods. The phenotypes do

Table 6.3. *Mortality and survival of the phenotypes in different periods*

(i) Numbers that survived or died
Period of observations

Phenotypes	1948–62		1973–75		1976–78	
	S	D	S	D	S	D
Pale	111	27	119	13	95	44
Intermediate	322	85	353	50	329	101
Dark	125	27	113	12	104	34
Total	558	139	585	75	528	179
Total in periods	697		660		707	

(ii) Analysis of χ^2

Component of variation	Value of χ^2	Dfs	Value of P
Differences in mortality between periods	43.6755	2	0
Differences in mortality between phenotypes	1.1611	2	0.560
Differences in number of phenotypes between periods	2.0920	4	0.719
Residual heterogeneity	4.5815	4	0.333
Total variation in $3 \times 3 \times 2$ table	51.5101	12	7.57×10^{-7}

In this table the column S consists of those that survived from one breeding season to the next; the column D consists of those that died.

not differ in mortality, nor do their numbers vary between the periods. The residual heterogeneity represents the component of variation determined by differences between periods in phenotypic variation: this is the interaction of the differences between periods with the differences between phenotypes; and this, too, is not significant.

Having found no differences in the phenotypic mortality rates, any selection we detect must depend on differences in age at maturity or reproductive rate. To calculate the fitnesses and selective coefficients, we must still choose a figure for the annual mortality. The mortality over all years 1948–78 is the mean of three completely different and heterogeneous values. The mortality after 1975 has been artificially raised by the human predation. Our choice must lie between the 19.94 per cent mortality for 1948–62 and the 11.36 per cent for 1973–75. I leave this problem to be discussed in section 6.3, where I present my demographic analysis of selection in the Arctic Skua population.

Table 6.4. *Distributions of age at maturity of the phenotypes*

(i) Numbers of birds reared in 1948–59

Phenotype	Age at first breeding (years)					Total	Mean age
	3	4	5	6	7		
Pale	6	5	1	2	—	14	3.929
Intermediate	—	16	8	2	1	27	4.556
Dark	1	5	3	2	—	11	4.545
Total	7	26	12	6	1	52	4.385

(ii) Numbers of birds reared in 1970–76

Phenotype	Age at first breeding (years)					Total	Mean age
	3	4	5	6	7		
Pale	3	7	4	—	—	14	4.071
Intermediate	3	10	10	3	—	26	4.5
Dark	—	4	4	1	—	9	4.667
Total	6	21	18	4	—	49	4.408

(iii) Combined numbers for 1948–59 and 1970–76

Phenotype	Age at first breeding (years)					Total	Mean age
	3	4	5	6	7		
Pale	9	12	5	2	—	28	4.0
Intermediate	3	26	18	5	1	53	4.528
Dark	1	9	7	3	—	20	4.6
Total	13	47	30	10	1	101	4.396

6.2 Age at maturity and reproductive rate in relation to age

Age at maturity is a part of the life history on which selection may act very strongly. The annual mortality of Great Tits is about 50% (Bulmer & Perrins, 1973). Suppose some Great Tits bred at one year old: others did not breed until two years old. Of those breeding at one year old, 50 per cent would survive to breed: of the others, only 25 per cent would survive to breed; they would suffer a 74 per cent selective disadvantage compared to the younger breeders who would thus spread throughout the population in a few generations. The enormous selective advantage of early sexual maturity is a consequence of the high annual mortality. Unless a tit breeds in its first year, it then has only a 50 per cent chance of breeding at all. Longer-lived birds mature at greater age. An extreme example is the Wandering Albatross, *Diomedea exulans*: its annual mortality is only about two per cent; it breeds for the first time at nine or more years old. An albatross breeding for the first time at eight years old would gain an advantage of only about two per cent. Of course, this small advantage is still an advantage. A gene that determined the earlier onset of maturity would still spread through the population. So we must ask: why doesn't earlier maturity also evolve in long-lived birds? Or in any bird that does not breed in its first year? To answer these questions, it is not sufficient merely to assume that a younger bird would be less successful raising chicks in its first year of breeding. Birds are usually less successful in their first year. In addition, and necessarily, earlier maturity must reduce their overall chances of breeding. Either they reduce their chance of survival and hence the length of their breeding life, or they are less successful parents than they would have been if they had started parenthood at a greater age. Earlier maturity must have a selective cost, probably of increased mortality, to balance its selective advantage.

This theory of a selective balance may be true but it is difficult to test. We cannot say how many chicks a bird would have produced if it had bred a year earlier, nor can we say how its chances of surviving to normal breeding age might have been reduced. But we can observe the age at maturity and reproductive rates at different ages among genetically different pheno-types – like the melanic and non-melanic Arctic Skuas. Indeed, this is precisely the sort of theory that can be tested by finding genetically marked differences. In a passerine bird, like the Great Tit, which breeds in its first year, there is, of course, nothing to be explained. The mortality is so high and the selective advantage of breeding in the first year so great, that no adverse cost is likely to be large enough to balance the advantage.

The melanic and non-melanic Arctic Skuas differ in both age at maturity and reproductive rate, but not in survival rate. Table 6.4 shows the distributions of age at maturity. These data are the ages when chicks, reared in the colony on Fair Isle, returned to breed. In both periods 1948–59 and 1970–76 (these being the years when the returning chicks were ringed, chicks having been ringed every year at the bird observatory), the ages at first breeding are very similar: the non-melanic pale birds breed at about four years old on average; the melanic dark and intermediate birds breed at about four and a half years old. The overall distributions of age at maturity are very similar in both periods:

Age at maturity	1948–59	1970–76
3	7	6
4	26	21
5	12	18
6 or 7	7	4

$$\chi_3^2 = 2.5402, P = 0.468$$

To analyse the differences between the phenotypes, the combined table of both periods must be condensed, as follows, to give large enough numbers in each class for the calculation of χ^2

	Age at first breeding (years)		
	3	4	5, 6 or 7
Pale	9	12	7
Intermediate	3	26	24
Dark	1	9	10

Then

$$\chi_4^2 = 13.7472, P = 0.00815$$

The four degrees of freedom on which this value of χ^2 is based can be analysed separately by dividing the table into four independent 2×2 contingency tables.

(i) Comparison between intermediate and dark at ages 3 or 4 years:

	Age at maturity	
	3	4
Intermediate	3	26
Dark	1	9

(ii) Comparison between intermediate and dark at ages ≤ 4 or ≥ 5:

	Age at maturity	
	≤ 4	≥ 5
Intermediate	29	24
Dark	10	10

(iii) Comparison between pale and melanic (intermediate or dark) at ages 3 or 4:

	Age at maturity	
	3	4
Pale	9	12
Melanic	4	35

(iv) Comparison between pale and melanic at ages ≤ 4 or ≥ 5:

	Age at maturity	
	≤ 4	≥ 5
Pale	21	7
Melanic	39	34

In this analysis of the 3×3 table, the numbers compared between two classes (for example intermediates and darks, or ages 3 and 4) are then summed to be compared with the third class (for example, intermediates plus darks compared with pales, or ages 3 plus 4 compared with ages 5 and older). Each 2×2 table thus makes a comparison that is independent of the others. The values of χ^2 that might be calculated by the standard formula for a 2×2 table are not exactly additive. Kimball (1954) devised a strictly additive method of calculating these values of χ^2 in this analysis of a contingency table. The overall expectations of the independent effects tested by each 2×2 table are calculated from the totals of the 3×3 table. A.E. Maxwell (1961) gives an account of Kimball's method. Table 6.5 shows the results of this analysis of χ^2.

The results of the analysis are clear: a greater proportion of pale birds breed for the first time at three years old than intermediate and dark melanics; a smaller proportion of pales breed first at five or six years old than the melanics. These are the only significant differences. Intermediates and darks are very similar in their distributions of age at maturity. So the difference is almost entirely between the non-melanic pales and the melanics.

Pales and melanics also differ in their reproductive rates. O'Donald & Davis (1975) gave a table showing the average number of chicks fledged by birds in each year from the first year they bred on Fair Isle. The data in this table are all derived from the period 1948–62, when the colony was smaller and the newly fledged chicks could more easily be assigned to particular pairs. In the period 1973–79, this was more difficult: it was necessary to observe each pair long and repeatedly at the time of fledging at the end of

Table 6.5. *Analysis of χ^2 of data on ages at maturity*

Component of variation	Value of χ^2	Dfs	Value of P
Difference in proportions of intermediates and darks breeding at ages 3 or 4 years	0.0019	1	0.965
Difference in proportions of intermediates and darks breeding at ages ≤ 4 or ≥ 5 years	0.1340	1	0.714
Difference in proportions of pales and melanics breeding at ages 3 or 4 years	9.7049	1	0.00184
Difference in proportions of pales and melanics breeding at ages ≤ 4 or ≥ 5 years	3.9064	1	0.0481
Total variation in the 3×3 table	13.7472	4	0.00815

the breeding season. After 1975, when I received only part-time assistance on the project, I made no attempt to observe the fledging of the chicks of all the pairs: I was concerned mainly with the problem of sexual selection and with obtaining data on breeding date and territory size. So the data on reproductive rates have all been obtained from the earlier period. Birds in their first year of breeding have a lower reproductive rate than in their later years. This is shown in Table 6.6, which gives the reproductive rates of individually known, ringed birds in their first and subsequent years of breeding. Table 6.6 has been condensed from table A.2 of Appendix A, in which the reproductive rates are shown for every year of breeding. Table A.2 reproduces the data in table 1 of O'Donald & Davis' paper on the demography of the Fair Isle population (O'Donald & Davis, 1975). More data become available if the reproductive rates of individual birds are counted for each pair of birds in the colony. The data of many unringed birds are thus included also. These data (taken from O'Donald *et al.*, 1974) are shown in table A.3 of appendix A. This gives the mean numbers of fledglings produced by birds that bred together as a pair for one, two or more years. Classification of the data by longevity of pairs obscures the effect of the lowered reproductive rates of birds in their first year of breeding; for many new, first-year pairs consist of experienced breeders with a new mate. Only a small proportion of new pairs consists of two new birds breeding for the first time. Thus although the data of birds in pairs show the overall differences between the phenotypes very significantly, they do not show the reproductive rates at different ages – only the reproductive rates for the years that the birds of a pair stayed together. For the demographic calculations of the next section, I have used the data in table 6.6 – data of known individuals.

Table 6.6. *Reproductive rates of ringed birds in their first and subsequent years of breeding*

Phenotype and breeding experience	Males			Females			Both sexes		
	n	Σx	$\bar{x}$	n	Σx	$\bar{x}$	n	Σx	$\bar{x}$
Pale									
First year	22	15	0.682	29	17	0.586	51	32	0.627
Later years	58	71	1.224	102	132	1.294	160	203	1.269
All years	80	86	1.075	131	149	1.137	211	235	1.114
Interm.									
First year	71	58	0.817	70	63	0.900	141	121	0.858
Later years	191	246	1.288	211	288	1.365	402	534	1.328
All years	262	304	1.160	281	351	1.249	543	655	1.206
Dark									
First year	32	28	0.875	24	23	0.958	56	51	0.911
Later years	108	161	1.491	64	85	1.328	172	246	1.430
All years	140	189	1.350	88	108	1.227	228	297	1.303
All phenotypes									
First year	125	101	0.808	123	103	0.837	248	204	0.823
Later years	357	478	1.339	377	505	1.339	734	983	1.339
All years	482	579	1.201	500	608	1.216	982	1187	1.209

In this table, n is the number of birds and Σx the total number of chicks they fledged.

The data from pairs agree with the data of individuals: pale, non-melanics differ from melanics in their mean reproductive rates. This is shown in the following table:

Numbers of chicks fledged per pair

	Data of ringed birds		Data from all pairs	
	No.	Mean	No.	Mean
Pale	211	1.1137	242	1.1157
Melanic	771	1.2348	1001	1.2038

The very small difference between pales and melanics arises largely from the relative lack of success of pales in their first year of breeding. The means of first-year breeding birds

$$\bar{x}_P = 32/51 = 0.627 \text{ (pales)}$$
$$\bar{x}_M = 172/197 = 0.873 \text{ (melanics)}$$

Table 6.7. *Reproductive rates of male and female phenotypes*

(i) Mean reproductive rates

Sex	Pale	Intermediate	Dark
Male	1.1080	1.1366	1.3465
Female	1.1214	1.2358	1.1685

(ii) Analysis of variance

Source of variation	Sum of squares	Dfs	Mean square	Value of F	Value of P
Between phenotypes	2.4436	2	1.2218	2.124	0.120
Between sexes	0.1078	1	0.1078	0.187	0.665
Interaction: difference of phenotypes between sexes	3.3866	2	1.6933	2.944	0.0531
Residual	617.729	1074	0.5752	—	—

This table is an analysis of the data on fledging success given in table A.3 of appendix A, which reproduces the data of O'Donald *et al.* (1974). The data are the numbers of chicks fledged by individual birds of each pair. They are divided into the number of years in which the birds had bred together as the same pair. The residual variance is the variance within phenotypes and years. The years in which the birds bred together as a pair have thus been treated for analysis as the 'blocks' in a randomized block experiment.

differ significantly as the following analysis of variance shows:

Source of variation	Sum of squares	Dfs	Mean square	Value of F	Value of P
Between phenotypes	2.4446	1	2.4446	4.177	0.042
Within phenotypes	143.9739	246	0.5853	—	—

O'Donald *et al.* (1974) showed that, although the three male phenotypes differed in their overall reproductive rates, the female phenotypes did not. Melanic and pale males differed very significantly. But this statistical significance is lost when the data of males and females are combined. Table 6.7 shows the mean reproductive rates and corresponding analysis of variance. In spite of the large numbers of observations, the only significant component is the difference between pales and melanics in their first year. This produces most of the overall difference in reproductive rates. But not all. Sexual selection also raises the reproductive rate of melanic males. In chapters 8, 9 and 10 of this book, I show how, when forming new pairs, females choose to mate with melanic males earlier in the breeding season. The melanic males then gain an advantage because earlier pairs fledge more chicks (see, particularly, sections 8.4 and 10.1). Dark melanic males gain the greatest advantage. This produces part of their increased reproductive rate. Although small in its overall effect, the sexual selection is statistically highly significant: selection acts differently on males and females.

The variation in the reproductive rates of Arctic Skuas appears to be determined by two factors: pales are less successful than melanics in their first year of breeding; in the formation of new pairs, sexual selection favours melanic males, thus raising their reproductive rate compared to pales. These results corroborate the theory that earlier maturity must be balanced by an overall reduction of reproductive rate. This has occurred, not by reducing the chances of survival and hence the expectation of life, but by reducing the breeding success of pale birds in their first year of breeding and the pale males' chances of mating early in the breeding season. The fitnesses and selective coefficients determined by these phenotypic differences can now be calculated.

6.3 Intrinsic rates of increase and components of selection

In appendix A, I have derived formulae for calculating intrinsic rates of increase in bird populations. I based my derivation on two assumptions: (i) after their first year of life, birds suffer a constant mortality in each subsequent year; and (ii) they produce a smaller than average number of chicks when they breed for the first time and then a constant number in every subsequent year they breed. We have already estimated the survival and reproductive rates from data given in the previous sections of this chapter. These estimates are, of course, mean values – means for each phenotype or for the entire population. The mortality rates do not vary between phenotypes. Differences in reproductive rate and age at maturity are shown in table 6.8. The values here have been taken from tables 6.5 and 6.6. However, the reproductive rates shown in table 6.6 are mean numbers of chicks fledged by pairs. In table 6.8, they have been divided by two to give the chicks fledged by individual birds.

Since mortality rates are closely alike for each phenotype, with no hint of any significant difference, I have assumed they are the same. Obviously, differences may exist that are too small to be detected statistically. If so, their selective effect would be correspondingly slight. But we still have to choose between the estimates of mortality

$$d = 0.1994 \text{ (for period 1948–62)}$$

and

$$d = 0.1136 \text{ (for period 1973–75)}$$

Table 6.8. *Variations between*
phenotypes in
reproductive rate
and age at maturity

Phenotype Reproductive rate	Age when first breeding (years)				
	3	4	5	6	7
Pale					
$b^* = 0.3137$	9	12	5	2	—
$b = 0.6344$					
Interm.					
$b^* = 0.4291$	3	26	18	5	1
$b = 0.6642$					
Dark					
$b^* = 0.4554$	1	9	7	3	—
$b = 0.7151$					
All phenotypes					
$b^* = 0.4113$	13	47	30	10	1
$b = 0.6696$					

In this table, b^* represents the reproductive rate in the first year of breeding, and b the average reproductive
rate in subsequent years of breeding.

Statistically, these estimates are completely different. There could be no
justification for taking the mean of both of them. The third estimate,

$$d = 0.2532 \text{ (for 1976–79)}$$

we have already rejected (section 6.1.1) because many birds had apparently
been shot. But how can we choose between the mortalities of the periods
1948–62 and 1973–75? Did shooting cause the higher mortality in 1948–62?
Or was it some other factor? The colony was smaller in 1948–62. It grew
with the growth of other seabird colonies – the Kittiwakes for example.
Whatever the factors for growth were, they were probably acting to reduce
mortality. But this is a vacuous speculation. All we can do is use each figure
for mortality in turn. Since we know that roughly 34.6 per cent of chicks
return to breed, we can estimate the first-year mortality, c, from the
subsequent annual mortality, d. Appendix A describes the method of
estimation. Thus we have the following estimates of mortality for

Table 6.9. *Intrinsic rates of increase and selective coefficients*

(i) Mortality as observed in 1948–62: $d = 0.1994$

Factors that determine components of selection

	Age at maturity		Reproductive rate		Both factors	
Phenotype	r	s	r	s	r	s
Pale	0.01987	0	−0.001190	0.141	0.009536	0.010
Intermediate	0.004388	0.123	0.008098	0.071	0.004251	0.053
Dark	0.002497	0.137	0.01675	0	0.01066	0

For all phenotypes, $r = 0.008225$ with generation time $T = 8.497$

(ii) Mortality as observed in 1973–75: $d = 0.1136$

Factors that determine components of selection

	Age at maturity		Reproductive rate		Both factors	
Phenotypes	r	s	r	s	r	s
Pale	0.07582	0	0.05635	0.170	0.06520	0.040
Intermediate	0.06225	0.129	0.06516	0.093	0.06195	0.071
Dark	0.06067	0.143	0.07469	0	0.06922	0

For all phenotypes, $r = 0.06546$ with generation time $T = 10.19$

calculating intrinsic rates of increase:

	First-year mortality	Constant mortality after first year
1948–62	$c = 0.2768$	$d = 0.1994$
1973–75	$c = 0.4817$	$d = 0.1136$

These values, and those in table 6.8, can now be inserted in the formulae for intrinsic rates of increase shown in appendix A. Table 6.9 then gives the estimated intrinsic rates of increase and selective coefficients corresponding to the factors that give rise to selection. The use of the lower mortality estimate, $d = 0.1136$, gives a more reasonable estimate of the rate of increase, $r = 0.06546$: this figure is roughly in agreement with the actual rate of increase of the Arctic Skua population on Fair Isle. From 1948–62, the population expanded from 15 to 71 pairs, at an average intrinsic rate

$$r = 0.109$$

Then from 71 pairs in 1962 to 106 pairs in 1973

$$r = 0.0364$$

Finally, from 1973–75

$$r = 0.110$$

The rate varies between periods with an overall average

$$r = 0.0694$$

As we should expect, the lower annual mortality gives the higher rate of increase. But although the different estimates of mortality produce different absolute rates of increase, the relative rates of increase of the phenotypes, and hence their selective coefficients, differ only slightly. For estimating the selection, the mortality is not a critical factor.

The results of these calculations show that the lowered reproductive rate of the pale phenotypes more than balances their advantage from earlier maturity. Natural selection thus favours the dark phenotypes. Intermediates are slightly more disadvantageous than pales. This heterozygous disadvantage should lead to the ultimate elimination of the pale phenotypes. Heterozygous disadvantage produces unstable points of equilibrium. A gene at a frequency greater than equilibrium will then spread through the population while a gene at a frequency below equilibrium will be eliminated. The selective values estimated in table 6.9 (ii) show that the point of unstable equilibrium occurs when the gene for pale is at frequency

$$p_* = 0.695$$

This is much greater than the actual frequency of about

$$p = 0.45$$

Thus pale phenotypes should eventually be eliminated from the population. But this calculation ignores the sexual selection. Since we know that sexual selection acts on the males, we can assume that the reproductive rates in table 6.6 reflect the sex difference. Intrinsic rates of increase and selective coefficients can thus be calculated for each sex, as shown in table 6.10. According to the estimated selective coefficients, the selection favours dark males and pale females. This difference between the sexes is partly caused by the sexual selection in favour of dark males.

When selection in one sex is reversed in the other, three points of equilibrium may exist, two of them stable and the other unstable (Owen, 1953). Given the estimates of table 6.10, the allele for pale should be at equilibrium at the frequencies

$$p = 1.0, 0.7068, \text{ and } 0$$

These equilibria include the two fixation states when the pale allele has either been fixed in the population ($p=1.0$) or eliminated ($p=0$). The fixation states are stable. The polymorphic equilibrium point ($p=0.7068$)

Table 6.10. *Selection within each sex*

(i) Reproductive rate and age at maturity

Phenotype	Reproductive rate Males	Females	Age when first-breeding (years) 3	4	5	6	7
Pale							
b*	0.3409	0.2931	9	12	5	2	—
b	0.6121	0.6471					
Interm.							
b*	0.4085	0.45	3	26	18	5	1
b	0.6440	0.6825					
Dark							
b*	0.4375	0.4792	1	9	7	3	—
b	0.7454	0.6641					

(ii) Intrinsic rates of increase and selective coefficients

Phenotype	Males r	s	Females r	s
Pale	0.06217	0.109	0.06692	0
Interm.	0.05815	0.145	0.06571	0.012
Dark	0.07351	0	0.06205	0.048

As in table 6.8, $b*$ represents first-year reproductive rate and b average reproductive rate in subsequent years; r is the intrinsic rate and s the selective coefficient. The calculations are based on the value $d=0.1136$ for the annual mortality.

is unstable: from either side of this point the allele moves to one of the fixation states. The unstable equilibrium divides the range of frequencies into two 'domains of attraction' to the fixation states. The condition

$$p \geq 0.7068$$

defines the domain of attraction to the point $p=1.0$: from all values of $p \geq 0.7068$, the pale allele increases in frequency to fixation. Similarly, the condition $p < 0.7068$ defines the domain of attraction to the point $p=0$. Since very nearly 20 per cent of the birds on Fair Isle are pales, the pale allele has a frequency of roughly 45 per cent ($p = \sqrt{0.2} = 0.45$). When we allow for the differences between the sexes, we again conclude that selection should eliminate the pale allele: its frequency lies within the domain of attraction to $p=0$. Its elimination will take a long time: it will take 100 generations (1000 years) to reduce its frequency to about one per cent.

Owen's theory of differential selection between the sexes was derived on the assumption that the selective coefficients are constants. As shown in chapter 10, however, it seems most probable that the sexual selection of the Arctic Skua is the consequence of preferential mating and therefore frequency-dependent. As the dark males increase in frequency, their sexual advantage should decline. If the sexual selection is the crucial component on which the overall advantage of dark depends, its frequency-dependent decline might produce a state of balanced polymorphism at some lower frequency of the pale allele. Immigration of pale birds would be required to balance the darks' advantage at their present frequency. After analysing the theory of sexual selection and its application to the Arctic Skua in chapters 9 and 10, I shall return to discuss the problem of the maintenance of the polymorphism in my concluding chapter.

7

Sexual behaviour

7.1 Territoriality, pair formation and mating behaviour

Observation of a few pairs of Arctic Skuas may reveal only very little of their courtship and mating behaviour. Pairs often copulate with no obvious preliminary courtship; they may have bred together for many years. Birds that know each other well enough perhaps have no need for elaborate courtship.

Perdeck (1963) observed the pairing of a few single birds and the mating behaviour of some established pairs in a colony of Arctic Skuas on the Faeroe island of Mykines. Since the birds had not been ringed, pairs from previous years could not have been distinguished from newly formed pairs. Single birds in the act of pairing may simply have been rejoining their former mates. If they had been seeking new mates, their behaviour might have been different. A detailed analysis of pairing and mating behaviour would have been possible on Fair Isle where all birds from previous years are known by their colour rings. But my own study of the population biology of the Fair Isle Arctic Skuas never allowed time for systematic observations of behaviour. Of course, I observed many aspects of their agonistic, pairing and mating behaviour. My own observations of their calls and postures agree closely with those of Perdeck, whose terminology I have used in the following account.

7.1.1 Agonistic behaviour in defence of territory

Perdeck (1963) observed the agonistic behaviour of Arctic Skuas defending their territories. Unlike their relatives, the gulls, skuas defend much larger areas round their nests. In a gull colony, nests may be packed closely together. Burger (1980), studying the territoriality of Herring Gulls (*Larus argentatus*), found that a territory with an average area of about 20 ± 5 m^2 was defended against all intruders. In the Fair Isle colony of Arctic Skuas, nests are never closer than 30 m. On average, about 100 m separate the nests. As we shall see later in this chapter (section 7.3), territories average about 11 000 m² in area. I shall argue (in section 7.4 of this chapter) that the territoriality of the Arctic Skua evolved partly by sexual selection.

A territory may also be a place for gathering food. A study of Arctic Skuas in Norway showed that isolated pairs had a more terrestrial diet, which presumably they had gathered in their territories, than colonial pairs, who presumably had much smaller territories to gather food in (Andersson & Götmark, 1980 – see section 3.4 for more details and discussion). This observation gives scant support to the theory that territoriality evolved to provide defendable areas to feed in: the colonial pairs, whose diet was almost entirely marine, still had very large territories, of average size 62 000 m² in one colony and 48 000 m² in another. The territoriality of the Arctic Skua must have evolved by some other selective advantage besides that of defending a feeding area.

Everyone who ventures into an Arctic Skua colony discovers almost immediately that the territories are defended fiercely. They are defended just as fiercely against other Arctic Skuas, Bonxies, sheep and dogs as against humans. The attack is from the air. The attacking skua dives at the intruder, sometimes striking with lowered feet. When another Arctic Skua is attacked, it crouches down in anticipation of the swoop, ducking its head at the last moment of the attack, then turning to watch the swooping bird climb away. Some very aggressive birds stand on an exposed spot while being attacked. They crouch at the beginning of the swoop, but counter-

A pale Arctic Skua starts to pull away at the bottom of its Swoop.

attack instead of ducking. They jump up as the swooping bird passes. According to Perdeck, photographs show that in this *Jump* the bird turns upside down in the air and strikes at the swooping bird with its feet. After the jump, the bird turns over and lands normally (see figure 7.1). Perdeck's description of this counter-attack certainly fits my own impression of it, but without photography the sequence of events cannot easily be followed.

After a certain number of swoops, the attacker lands back in his territory. The intruder may then attempt to swoop at the territory holder. This may produce long *Pursuit Flights* in which the birds chase each other at high speed. They often fly far from the territory of the original attacker,

An Arctic Skua stands 'at ease'. Another makes the Long Call.

Figure 7.1 The 'Jump' in counter-attack to the 'Swoop'.

attempting to strike each other while in flight. If one bird in the chase is attacking from above, the lower bird may turn upside down, as in the jump, in an attempt to strike the other.

As well as aerial combat, calls and aggressive postures are used to intimidate rivals. Perdeck describes a *Long Call* used to defend territory:

'It is composed of a series of loud, clearly, bisyllabic and rhythmically repeated notes ('gee joo'). The number of notes within one call varies from 1 to 12, but in most cases 3 or 4 notes are given. One note lasts about 3/4 second and there is no pause between two notes. The last note from a series is often incomplete. The Long Call is heard from sitting or standing birds, as well as during flight. During the call, the neck of a standing bird is mostly somewhat stretched and at each note the head is moved slightly up and down. The bird called at is followed with one eye, frequently first one eye and then the other alternately, so that the head moves from side to side'.

I would describe the long call as 'eeeow eeow'. This wild, rather melodious, screaming cry will always remain as one of my chief memories of the Arctic Skuas – so well in tune with the wild and beautiful places where they breed. It forms the continual background music of a skua colony. Birds sitting on the nest, or just standing or sitting in their territories, call at birds passing overhead. At the end of an aerial pursuit, the pursuer and the pursued call as they break off and return to their territories.

During pursuit flights, a different call, known as the *Short Call*, is often repeated, especially when attacking or just after landing. It is similar to the staccato alarm call of gulls. Perdeck also heard another call, the *Yelp*:

'It is a high, piercing sound, resembling the yelp of a dog. It is not rhythmically repeated, although a series may be produced. Sometimes it is bisyllabic, pointing to a relation with the Long Call. It was heard from birds being swooped at and after real chases'.

Having sexed some of the birds by watching pairs copulate, Perdeck observed the calls made by each sex:

'An interesting observation on the calls was made when a pair was being swooped at by a neighbouring bird. Both male and female often Long Called, but in addition to this the female gave Yelps, the male Short Calls. Further, while the male fought back with Jumps, the female never did and finally hid in a hole. This observation, together with others, suggests that the Yelp indicates a stronger escape tendency, the Short Call a stronger aggressive tendency, with the Long Call as an intermediate'.

Observations of many pairs would be necessary to establish how widely this theory may apply. Both birds of a pair are often observed making long calls. I never attempted to note the sex of birds making yelps or short calls or their reactions at the time.

Two postures are characteristic of confrontations on the ground. An Arctic Skua that walks into another's territory is usually driven away by the long call combined with the *Aggressive Upright* posture (figure 7.2). In this posture, the bird stands upright with neck stretched, head held high, and bill pointing horizontally or downwards. This posture is very intimidating. The intimidation is reinforced by walking towards the intruder or by long calling (Perdeck, 1963). After this display, the intruder is often pecked. When threatened, an intruding bird often adopts the *Intimidated Upright* posture (figure 7.2), which Perdeck describes as follows:

'The bill points obliquely upwards and the neck is held more or less backwards. The stretching of the neck is the same as in the Aggressive Upright. A bird moving from the Aggressive into the Intimidated Upright gives a queer stiff impression, with the neck drawn back and the whole body tilted correspondingly as if it were rigidly fixed to the neck. The immobility of the posture is quite characteristic in a bird that otherwise is so very active. The owner of the territory sometimes walks round the intruder (especially during pair formation conditions). As the owner passes behind him, the intruder, standing in the Intimidated Upright, may very quickly turn his head only towards the territory holder, probably at the moment the latter would otherwise move out of sight. The immobility, and probably the posture itself too, seems to have the function of avoiding the arousal of aggression in the other bird as far as possible'.

7.1.2 Pair formation and mate selection

Perdeck (1963) observed two single males on territories. They defended their territories strongly, though not as strongly as mated pairs.

Figure 7.2 Aggressive and intimidated upright postures.

He recognized two females who visited several territories, not only those of the single males but also those of established pairs. He says:

'This suggests that during pair formation the males establish a territory and that the females search for such birds. A number of meetings between these unpaired females and a single male in his territory was observed. There was no clear Meeting Ceremony as is described in the gulls. The female tried to stay in the territory, frequently flying up and returning after a short flight around. When she approached the male he assumed the Aggressive Upright and often walked around the female. She also stood in Upright, but mostly in the Intimidated Upright. Perhaps this posture serves as an appeasement gesture, comparable with the Head Flagging of the gulls in the same situation. Already in this early stage, the male tries to copulate, though this is rarely tolerated by the female. Often the male introduces a Squeaking Scene or a mixture between this and a precopulatory display, followed by nest building movements'.

(A *Squeaking Scene*, described in (iii) following this section, determines the nest site.)

Perdeck's observations of these few birds suggest that the males secure and defend the territories. The females visit them in turn and choose between them. In models of the evolution of territoriality (section 7.4 of this chapter) and models of female choice of male phenotype (section 10.1 of chapter 10), I have assumed that the unmated males are all available on their territories. As the females, either successively or in groups, reach a state of readiness to breed, they seek their mates from among the pool of unmated males. This is, indeed, one of the premises of Darwin's theory of sexual selection in monogamous birds (see section 8.1, chapter 8).

If the male has tenure of the territory, having originally claimed and defended it, we should expect that when pairs separate, the male would stay put while the female would seek a new mate elsewhere. I have observed males defending their territories of the previous year while waiting for their mates to return. If they do not return within a few days, a new mate may be found. Sometimes the female of a pair returns to find no mate awaiting her. She may then defend the territory herself and take a new mate on her former territory. Many pairs are lost simply because one of the pair has died in the winter or on migration. Some are lost on Fair Isle because they are illegally shot by islanders soon after their return. Some pairs divorce, and each divorcee then finds a new mate. These divorces may occur if one bird returns to the territory much later than the other. The first bird to come back soon tires of just sitting in its territory and tries to join adjacent pairs. Threesomes can thus form early in the season. The established pair frequently adopts the aggressive upright posture towards the new bird, who in turn adopts the intimidated upright posture. The new bird is quickly

Table 7.1. *Change of position of territory after change of mate*

(i) Position of present territory relative to former territory held with previous mate

Category of change in position of territories[a]

	(0)	(1)	(2)	(3)
Males	21	9	1	2
Females	10	7	14	12

[a] (0) signifies no change: the new pair are breeding on the same territory, or on a territory that largely overlaps, the former territory; (1) signifies that the new pair have a territory adjacent to, or only slightly overlapping, the former territory; (2) signifies that the new pair's territory is separated from the former territory by another pair's territory or by a large area of unoccupied ground; and (3) signifies a territory widely separated from the former territory, often in a different part of the colony.

(ii) Analysis of χ^2

Comparison between categories	Value of χ^2	Dfs	Value of P
Between categories (0) and (1)	0.5673	1	0.451
Between categories (2) and (3)	0.1711	1	0.679
Between categories (0)+(1) and (2)+(3)	20.8829	1	4.88×10^{-5}
Between all categories	21.6213	3	7.82×10^{-5}

made to feel unwelcome and usually driven off. But some birds are more persistent. The established pair may separate with the formation of a new pair. This in turn leaves another divorcee who may form a threesome with another pair. Further changes of mate may thus follow. As a result, males too may sometimes change their territories on changing their mates. But much more often, it is the female who moves to a new territory with a new mate.

To show that on changing mates males stay put whereas females often move, I classified four categories of change of position of territory from one year to the next. In *Category (0)* the territory either remains the same or largely overlaps the former territory. Of course, no territory is ever exactly the same from one year to the next, but the overlap may comprise much the largest part of the two territories. In *Category (1)* the territory in one year is adjacent to or slightly overlaps the territory in the former year. The change in position in *Category (2)* involves a jump across a third territory or across unoccupied ground: the territory in one year is separated from the former territory by another pair's territory or a large area of unoccupied ground (larger than an average sized territory). In *Category (3)*, the territories in successive years are widely separated, often in different parts of the colony. The data of new pairs in 1979 and 1978 are shown in table 7.1 in relation to these categories of how far they moved after their change of mate. The main distinction lies between categories (0) and (1) on the one hand, and categories (2) and (3) on the other.

Males very seldom move further away than to an adjacent or overlapping territory. Females often move to a completely different part of the colony. When category (0) is combined with (1) and (2) with (3), we obtain the 2×2 table

Move on changing mate

	To an overlapping or adjacent territory	To a territory elsewhere
Males	30	3
Females	17	26

Then $\chi^2 = 20.8829$ as shown in table 7.1. Yates' correction may be applied for greater accuracy in this case, giving

$$\chi^2 = 18.7625$$

corresponding to

$$P = 1.48 \times 10^{-5}$$

These observations show that the male has tenure of the territory. The female may retain it if the male dies. Otherwise, after divorce, she moves elsewhere.

New pairs are more likely to separate than pairs who have previously bred together. Both new and old pairs are equally likely to be lost by the death of one of the partners. New pairs are more often lost by divorce. The proportions of pairs that are lost in the following year on Fair Isle are shown in table 7.2. About 45 per cent of new pairs are lost compared to 35 per cent of old pairs. The numbers are compared in the following table:

	Pairs remaining together	Pairs lost
New pairs	134	111
Old pairs	249	137

As shown in table 7.2, when losses of new and old pairs are compared, we obtain

$$\chi^2 = 6.0508$$

When Yates' correction is applied to the calculation,

$$\chi^2 = 5.6464$$

corresponding to

$$P = 0.0175$$

Since Yates' correction only applies to the calculation of χ^2 from a 2×2 table, it cannot be applied to the orthogonal analyses of the χ^2s in tables 7.1 and 7.2. These calculations of χ^2 show that the difference between the proportionate losses of new and old pairs is significant, though not very great.

The proportionate losses of all pairs also vary significantly between years. The losses increase in later years:

Year	Percentage of pairs lost
1974	28
1975	35
1976	39
1977	46
1978	47

This must be a reflection of an increase in mortality by shooting. After 1976, I had no full-time research assistant working on Fair Isle throughout the breeding season. I could only arrive myself at the beginning of June, after the University examinations. So in 1977 and 1978, the Arctic Skuas were no longer under continual surveillance from the time of their arrival in late April and early May. A number of adults were apparently shot in this period in 1977 and 1978, since we found some of the bodies. The increase in losses of individual birds and pairs was the consequence of this human predation. For this reason, as I explained in chapter 6, section 6.1, the data on survival rates for the period of 1976–79 have not been used in the demographic calculations of population growth and selection. For estimating the mating preferences according to the models of sexual selection and assortative mating (chapter 10, sections 10.2 and 10.3), the shooting merely added to the number of new pairs that would be formed in the following year, and hence added to the available data on new pairs.

7.1.3 *Mating behaviour and nest building*

In old pairs who bred together in previous years, mating often occurs with little or no preliminary courtship or display. Perdeck (1963) describes copulation in these words:

'The male approaches the female in the Aggressive Upright, sometimes walking around her, while she assumes the same posture. If the female is lying down he pecks

Table 7.2. *Fidelity of new and established pairs*

(i) Numbers of pairs remaining together or lost in subsequent year

Year	Newly formed pairs			Old pairs established in previous years		
	Remain together	Lost	Total	Remain together	Lost	Total
1974	32	19	51	51	14	65
1975	32	22	54	57	26	83
1976	32	19	51	48	33	81
1977	25	26	51	46	35	81
1978	13	25	38	47	29	76
Total	134	111	245	249	137	386
Proportion lost $p=0.453$				$p=0.355$		

(ii) Analysis of χ^2

Component of variation	Value of χ^2	Dfs	Value of P
Difference in proportions of losses of new and old pairs	6.0508	1	0.0139
Differences between years in losses of all pairs	12.5251	4	0.0138
Differences between years in proportions of new and old pairs	2.7622	4	0.598
Differences in different years in proportions of losses of new and old pairs	4.8621	4	0.302
Total variation in the $2 \times 2 \times 4$ table	26.2002	13	0.0160

her gently at the wingtips or back. Often, before mounting, the male makes some quick movements with the neck, which look like intention movements of pecking at the female as he does when mounted. Meanwhile the male utters a call, the *Copulation Call*. It is a high-pitched, rhythmic sound, not unlike the barking of a small dog. The female assumes the *Willing Attitude*; a posture similar to the Hunched of gulls: the body is held horizontal, the neck is drawn in and the head is held at the same level as the body. In most cases the carpal joints are held out a small distance from the body and the tail is not raised. When mounting, the male calls more loudly and more quickly, and keeps his balance with flapping wings. There are usually 3–5 cloacal contacts, during which the male rests with his wingtips on the ground. Especially during the contacts the female makes a soft sound (*Begging Call*). This call varies a good deal: we heard a soft purr and a not-rhythmically-repeated 'kike'. Sometimes neither male nor female call during the copulation'.

Elaborate displays may precede copulation. Aerobatic flying displays take place (Witherby *et al.*, 1941). But they are very difficult to follow. *Begging* is very common. I have often seen this display from hides while attempting to catch birds on their nests in clap-nets. The male lowers his neck, indicating his willingness to regurgitate. The female pecks at the male's bill and he pecks at hers; both may open their bills widely. Often the female makes upward thrusts with her bill, trying to peck into the male's mouth. He soon regurgitates and she takes the food. Sometimes they copulate soon after the female has fed. In pairs who are well established and have already copulated, the begging display may have no other function than that of simply feeding the female. Copulations may take place without begging or any other apparent display. Figure 7.3 shows Begging and Squeaking. Figure 7.4 shows the Willing Attitude of the female and copulation.

Figure 7.3 The 'Begging' and 'Squeaking' attitudes.

Nest building is preceded by a characteristic call and display which determines where the nest is to be built. Perdeck identified a *Nest Call*, which is a nasal, bisyllabic sound ('hé-hé'), and used by the male as he flies to a suitable area, lands and makes scraping, nest-building movements. The female may then join the male, making the nest call. This may then become the *Squeaking Call* which leads to a *Squeaking Scene*, which Perdeck describes as follows:

'When the female is present, she may follow the male, and, when standing together, the Nest Call of both birds fuses into the *Squeaking Call*. This Call is also mostly bisyllabic. The first part is a harsh hissing squeak, the second part lacks the squeak, but the hissing remains. The male stands usually with a slightly lowered, fairly drawn-in neck, the bill pointing more or less at the ground. At times he bends

Figure 7.4 Willing attitude of female and copulation.

towards the female with a widely opened bill (calling). The female holds her head lower than that of the male, often stretching her neck and pecking upwards or sideways to the bill of the male. He also pecks at her bill and then the postures of the birds can be quite the same. This may go on for a longer time, and meanwhile nestbuilding movements are performed, especially by the female (Scraping, Sideways Nest Building). It was observed, that, by repeating this *Squeaking Scene* the ultimate nest site was determined. In the few cases studied, it was the male that chose the site. The real building of the nest is done mainly by the female, since she makes the most intensive nestbuilding movements'.

The nest itself is just a scrape in the ground. On Fair Isle, in the drier parts, the nest is usually a scrape in short-cropped heather or grass. It may be lined with thin heather twigs or short straws. In soft, boggy areas, a more obvious depression is made, more like the deep scrape of a Bonxie, and more material is used to line the nest. Some pairs make no attempt to produce a scrape or nest. One pair produced eggs in the middle of a narrow sheep walk across the heather. No scrape was made or nest lining used. Another pair's eggs were laid on a patch of small stones between two larger rocks. Where the heather has been grazed to an almost flat, tight lawn, the nest scrape is often minimal and the eggs stand exposed on the top of the heather. On Fair Isle as we have seen, the peat was cut right down to the rocky subsoil. Heavy grazing has kept the moorland vegetation thin and sparse. Good nest sites are not to be found in some pairs' territories.

7.2 Measuring territory size

A territory is a defended space. To measure its area, its boundary must be mapped. Points must be determined at which an intruder is first attacked. For example, an ornithologist could penetrate an Arctic Skua's territory from various directions and note the points where an attack is first launched. By surveying these points, a map could be drawn of the boundary line that passes through them. But this line will not necessarily define a constant and unique area. Territory size will vary with the holder's aggressive intentions. These will vary with his hormonal levels and state of well being. Territories often vary in size at different times in the breeding season. Different areas may be defended against different intruders. Burger (1980) observed that pairs of Herring Gulls defended three different areas: a primary territory, a secondary territory and a unique territory. A pair defends its primary territory against all conspecific intruders. But an incubating bird on its own only defends a smaller, unique territory. Thus the primary territory is defended if both birds are present, whereas the unique territory is always defended. A pair also defends a larger, secondary territory against non-neighbouring birds but only when their neighbours are not attacking the intruding non-neighbour. The primary and secondary territories overlap those of neighbouring pairs. The unique territories did not overlap each other, hence their uniqueness. A defended area is not a simple concept, always defined by a constant boundary.

Arctic Skuas attack different intruders at different times in the breeding season. A male Arctic Skua defends its territory against other skuas (Arctic Skuas and Bonxies) soon after it has returned in April or May. But human intruders are not attacked until the first egg has been laid. Sometimes a pair will make a few half-hearted swoops a day or two before their first egg appears. These swoops are made at slow speed and broken off early, well before reaching the target. Injury feigning always indicates that a pair has an egg. They then swoop at higher speeds and press home their attacks more closely. When the clutch is complete they often strike with lowered feet, just as they strike other skuas on their territory. They seem to attack humans most fiercely when their eggs are hatching. Gradually, as the chicks get nearer to fledging, their ferocity lessens. Attacks on humans cease once the chicks are flying. But Bonxies are still attacked just as furiously at the end as at the beginning of the breeding season. These variations in agonistic behaviour towards different intruders are obviously adaptive. Humans, sheep and dogs are threats only when a clutch has been laid or chicks are wandering about in the territory. Other Arctic Skuas or Bonxies deprive a male of his opportunity to breed by depriving him of his territory. As we saw in chapter 2, section 2.3, Bonxies have a formidable advantage in the competition for territories. They arrive back on the breeding grounds earlier in the season than the Arctic Skuas, who cannot dislodge them from a territory they have claimed. Towards the end of the breeding season, the slow and clumsy flights of the Arctic Skua chicks make them the easy targets of Bonxies on nearby territories. Many Arctic Skua chicks are lost to Bonxies on Foula. So the Arctic Skuas continue to defend their territories against Bonxies until the end of the breeding season when no other threats remain. An Arctic Skua's defence of his territory is not the same against all intruders at all times. His territory size will partly depend on the type of intruder and the stage in the breeding cycle.

How then can territory size be measured? Can estimates be obtained that are comparable between pairs? It would be very difficult to determine by observation at what points between nests neighbouring pairs of Arctic Skuas attack each other. The attacks are usually aerial: a dog-fight takes place. The combatants often violate the air space of other pairs, who then attack. Quickly the battle escalates. After screaming high-speed pursuits, pairs gradually leave the battle as it moves away from their own air-space.

Soon all is quiet again. In the midst of the melée, how could you ascertain a point on the ground where a particular pair had joined the battle? In the middle of the breeding season, when territories are strongly defended against all comers, you could note points where you are yourself attacked on entering a territory. This would be less subjective than attempting to note points at which one pair of birds attacks another. It would only be valid if the territory of each pair were determined at the same stage in the breeding cycle – an almost impossible requirement. It would only be practical with an army of assistants working full-time.

Fortunately, territory size can be estimated objectively using a method that depends on the theoretical consequences of the fact that the Arctic Skuas' territories share common boundaries. When territories share common boundaries, they necessarily become polygonal. This is dramatically illustrated in E.O. Wilson's book *Sociobiology* (1975) showing polygonal territories of fish revealed by polygonal patterns in the sand of a pool. This illustration was taken from Barlow's paper on the territories of the mouthbrooder fish, *Tilapia mossambica* (Barlow, 1974). Neighbouring Arctic Skuas certainly share common boundaries of their territories. If you walk across a boundary from one territory to another in the middle of the season when all pairs have eggs or chicks, the pair of the territory you are leaving will continue to attack until the pair of the territory you are entering start to feign injury. In the middle of the colony there is very little 'dead' ground where intruders are not attacked. This is demonstrated by attacks on sheep who have strayed into the Arctic Skua colony. Except in one or two small patches of dead ground they are continually attacked and driven backwards and forwards. Once in a patch of dead ground, they cannot escape, for they are immediately attacked and driven back when they try to leave. Normally, sheep are not found in the breeding season in the skua colony. But when they are rounded up from the north and west of Fair Isle for shearing, they are taken to a sheep dip near Vaadal at the south-east edge of the skua colony. They are released there after shearing and make their way back across the hill. As they pass through the skua colony, they are driven to and fro, from one territory to another, until, by a kind of random walk, they have reached their grazing areas. In their erratic course, they may enter a dead area between territories, where they are pinned down. As more sheep enter the area, they huddle together unable to move. For two years a dead area existed in Homisdale below the airstrip. In 1979, about 30 sheep were trapped here in a curious triangular-shaped huddle. The moment one of the sheep tried to move away, it was attacked. At night, the sheep can wander away without being instantly attacked. After two days the huddle had dispersed. Two other areas of dead ground existed where small huddles of sheep collected. The occasional existence of areas of dead ground shows that most territories are contiguous.

When moving in a line from one nest to another, we must cross the common boundary of the territories. The two pairs will defend their territories up to this point. The boundary will of course extend on either side of the line joining the two nests. We may assume the boundary will extend at right angles to the line between the nests. At some point, this boundary will cut other boundaries between these nests and other neighbouring nests. Each nest will thus be surrounded by a polygonal territory as shown in figure 7.5. Prior to Barlow's observation of hexagonal territories of the mouthbrooder fish, Grant (1968) had shown that Pectoral Sandpipers, *Calidris melanotos*, formed polygonal territories at high population densities. High density ensures, of course, that territories share common boundaries and thus become polygonal. Davis & O'Donald (1976b) fitted polygons round maps of the Arctic Skuas' nests on Fair Isle. Assuming that the pressure of two adjacent pairs balances on the perpendicular bisector of the line joining their nests, this line is the common boundary between their territories. The common boundaries with other adjacent pairs meet to produce the vertices of each polygon. The polygons thus produced are known as Voronoi polygons. Hasegawa & Tanemura (1976) fitted a model of the formation of Voronoi polygons to Barlow's data of the polygonal territories of the mouthbrooder. Various tests of the goodness of fit of the model gave no significant deviation: an adequate fit was obtained.

The boundary between adjacent territories could be found more accurately by estimating the point on the line between the nests where defence of one territory begins and the other ends. The boundary could then be drawn through this point at right angles to the line between the nests. A more aggressive pair would shift their boundary from the bisector of the line towards the nest of the more timid pair. By putting the boundary along the bisector, the differences between the territories may thus be underestimated. But without an army of assistants to help me, I chose the method of fitting Voronoi polygons to a map of the nests. Since this method cannot overestimate differences between territory sizes, it should not produce erroneous inferences about any real differences though it may not always be sufficient to detect them.

The mapping was carried out using a Brunton compass and a 100-m tape. Bearings and distances were taken from each nest to as many nests as possible. This produced much redundant information, but gave independent checks of the positions of the nests. The map was remarkably accurate. For example, the map was drawn by starting with the nests at one end of the colony, say at Wirvie Brecks, proceeding to map nests along the Brae of Restensgeo, across Eas Brecks, Homisdale and on to Byerwall, Sukka Mire and Brunt Brae; at the same time proceeding from the nests on Wirvie and Brae across Swey (see figure 1.1, chapter 1). The end of the map on Swey could then be joined with the end on Brunt Brae. The error, having thus proceeded in a rough circle, was never more than two metres when compared with the actual measured distance of the nests at the two ends. Lines were drawn between the adjacent pairs of nests and their perpendicular bisectors found and drawn by ruler and compasses. The intersections of these bisectors are then the vertices of the Voronoi polygons. The maps of the nests and territories were drawn on large sheets of graph paper on which the areas could easily be determined by counting squares. Figures 7.5 and 7.6 show maps of the territories on Wirvie Brecks in 1977 and 1978.

Figure 7.5 Map of polygonal territories on Wirvie Brecks and Brae of Restensgeo in 1977. The nest positions are shown with the pair number alongside. A Bonxie's nest is indicated by the letter B.

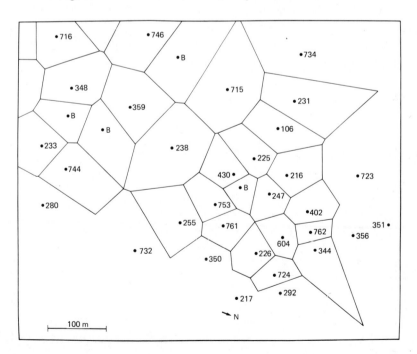

7.3 Territory size, phenotype and breeding time

In a territorial species, females choose only to mate with males on territories. Inevitably, therefore, sexual selection strongly favours male territoriality. Natural selection may act just as strongly as sexual selection. Male Red Grouse (*Lagopus lagopus scoticus*) without a territory can neither obtain a mate nor obtain sufficient food for survival (Watson & Moss, 1971). If space is limited, a male always gains an advantage simply by holding a larger than average territory: the more space he occupies, the greater is the number of males who cannot find territories and who are thus denied food and sex. Limitation of resource gives rise to competition and hence to selection (see section 7.4).

Territoriality is one of the functions of agonistic behaviour. More aggressive birds have larger territories. The male sex hormone, testoster-one, is a determinant of agonistic behaviour, and higher levels of testosterone increase territory size. Watson (1970) found that in Red Grouse, implants of testosterone increase the aggressive component of behaviour and territory size. Unsuccessful male grouse, without territories,

Figure 7.6 Map of polygonal territories on Wirvie Brecks and Brae of Restensgeo in 1978.

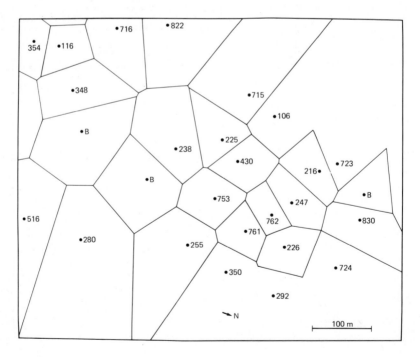

were found to establish territories after a testosterone implant. They could then feed, for the territory is a feeding area, and later obtain a mate, for the territory is also an area in which to mate and rear offspring. Genetic differences in aggressive behaviour will thus be subject to selection. A gene which raises the testosterone level will pleiotropically increase aggressiveness and hence territory size. It will be selected if the advantage of having a larger territory is not outweighed by some other disadvantage – for example the disadvantage of spending too much energy defending the territory and too little feeding or protecting the young.

Melanism is closely correlated with reproductive success in pigeons (Murton *et al.*, 1973). The melanic morphs, common in towns, have a longer breeding season and develop larger testes than the typical non-melanic morph. Many melanics remain in breeding condition even when kept in a 24-h cycle with only eight hours of daylight. The differences between melanics and non-melanics must be caused by differences in the capacity to secrete gonadotrophin. If melanics have a lowered threshold of photoresponse, they would produce gonadotrophin at shorter daylengths. Their gonads would recrudesce earlier in the breeding season, grow to a larger size, secrete higher levels of testosterone, and regress later. The melanics would thus become sexually more active and aggressive than the non-melanics. Murton *et al.*'s data of the frequencies of matings between the different phenotypes show that melanics are mated preferentially. Davis & O'Donald (1976*b*) fitted a model of preferential mating to these data: the melanics were the most preferred phenotypes. The melanics also produced a greater average number of offspring.

Pigeons are not conspicuously territorial. If they were, we should expect that the sexually more active melanics would maintain the larger territories. Davis & O'Donald (1976*b*) attempted to test this expectation by comparing the territories of melanic and non-melanic Arctic Skuas. Do melanics have larger territories? If so, does this give them a sexual advantage? The Arctic Skuas' territories were mapped on Fair Isle in 1974 and 1975. Pairs returning to their old territories were not included in the analysis, since pairing had taken place in a previous year. The new pairings, however, might have been determined by the choices of the females and the territory sizes of the males. Analysing the territory sizes of new pairs, Davis & O'Donald (1976*b*) found that territory size varied between the melanic and non-melanic phenotypes and was correlated with breeding date. The pale, non-melanic males had the smallest territories; pairs with smaller territories bred on average somewhat later in the season than pairs with larger territories. These effects are statistically significant only when experienced males, who had bred in previous years, changed their mate. Data of males breeding for the first time in the colony gave no significant correlation of territory size with breeding date.

If females searched at random for unmated males on their territories, a larger territory would be of direct sexual advantage to a male, for females would more often enter larger territories than smaller ones. Males on larger territories would obtain mates earlier in the breeding season. Since early breeding is correlated with higher reproductive success, then as Darwin postulated, sexual selection would operate in favour of the males whose characteristics – such as their aggressiveness in holding larger territories – gave them a higher chance of mating. This is a particular example of Darwin's general theory of sexual selection, which I discuss in section 8.1 of the next chapter. In sections 8.4 and 8.5, I show how the phenotypes are sexually selected in Arctic Skua populations. If the melanic males are the more aggressive of the phenotypes and do indeed hold the larger territories, this would explain their earlier breeding dates when they form new pairs. As we should expect, old pairs showed no phenotypic difference in territory size or breeding date: sexual selection depends on variation in the chances of finding a mate; only new pairs show its effects.

Davis & O'Donald's data for the years 1974 and 1975 consisted of 32 estimates of territory size for new males who were breeding for the first time and 35 estimates for old, experienced males in new pairings. These data are fewer than the total number of new pairs. Not all new pairs can be allocated territories: a Voronoi polygon cannot be constructed round a nest on the edge of the colony, since one side of the territory will not be bounded by other territories. Some new pairs have not been sexed as individuals. Breeding dates are only available for pairs who succeeded in hatching their young. In view of the small samples of territory sizes obtained in 1974 and 1975, I repeated the mapping of the nests in 1977 and 1978. This gave a further 47 estimates of territory sizes of males in new pairs. Breeding dates are known for 34 of these pairs. Table 7.3 shows the mean territory sizes of melanic and non-melanic males and gives the analysis of variance of the 114 estimates of territory size (67 estimates for the years 1974 and 1975 and 47 estimates for the years 1977 and 1978). The analysis of variance gives no hint of any significant effect of phenotype on territory size. The mean squares for both phenotype and male experience are less than the residual mean square. This analysis therefore contradicts Davis & O'Donald's conclusions based on the data for 1974 and 1975. They found that territory size differed between the phenotypes of old males with previous breeding experience. These data give the following analysis of variance:

Table 7.3. *Territory sizes of new and experienced males in new pairs*

(i) Mean territory size (m^2)

	Phenotype			
	Dark	Intermediate	Pale	All phenotypes
New males	10 437	10 065	11 732	10 489
Old, experienced males	14 778	11 345	9772	11 483

New males are males new to the colony on Fair Isle.
Old, experienced males have bred in previous years and have changed their mates, forming new pairs.

(ii) Analysis of variance in territory size

Source of variation	Sum of squares	Dfs	Mean square	Value of F	Value of P
Between phenotypes	53 525 490	2	26 762 745	0.5604	0.573
Between new and old males	29 441 000	1	29 441 000	0.6165	0.434
Interaction	101 005 449	2	50 502 725	1.0575	0.351
Residual	5 157 869 830	108	47 758 054	—	—

The interaction mean square measures the variation between phenotypes in the differences of new and old males.

Source of variation	Sum of squares	Dfs	Mean square	Value of F	Value of P
Between phenotypes	140 162 727	2	70 081 364	3.740	0.0347
Residual	599 637 381	32	18 738 668	—	—

The mean territory sizes were then

$$\bar{x}_D = 11507 \text{ (darks)}$$
$$\bar{x}_I = 11751 \text{ (intermediates)}$$
$$\bar{x}_P = 6693 \text{ (pales)}$$

This significant variation between phenotypes disappears when the estimates of territory size in 1977 and 1978 are included in the analysis of variance of old males with previous experience:

Source of variation	Sum of squares	Dfs	Mean square	Value of F	Value of P
Between phenotypes	122 776 529	2	61 388 264	1.4292	0.248
Residual	2 534 200 446	59	42 952 550	—	—

The data of new males reduces the differences between phenotypes still further, as shown in table 7.3. When new and old males are compared, their

Table 7.4. *Territory sizes of males in periods 1974–75 and 1977–78*

(i) Mean territory size (m²)

	Phenotype			
Period	Dark	Intermediate	Pale	All phenotypes
1974–75	11 991	10 407	8 152	10 162
1977–78	12 956	11 307	15 505	12 267

(ii) Analysis of variance in territory size

Source of variation	Sum of squares	Dfs	Mean square	Value of F	Value of P
Between phenotypes	43 500 070	2	21 750 035	0.4705	0.626
Between periods	176 638 460	1	176 638 460	3.8213	0.0532
Interaction	175 340 166	2	87 670 083	1.8966	0.155
Residual	4 992 253 730	108	46 224 572	—	—

means give rise to the following mean differences:

Phenotype	New Males	Old Males	Mean difference of new and old males
Dark	10 437	14 778	−4341
Intermediate	10 065	11 345	−1280
Pale	11 732	9772	1960

The interaction mean square measures the variance in the mean difference of new and old males. This, too, is not significant. Davis & O'Donald's earlier significant result may be explained either as the occurrence of a rare event by chance, or as the result of differences in the effects of territory size on mating in the two periods 1974–1975 and 1977–78. Table 7.4 shows an analysis of the data when classified by phenotype and period (1974–75 or 1977–78). The mean territory size of all phenotypes increased considerably between the two periods – from 10 162 in 1974–75 to 12 267 in 1977–78 – an almost significant change, with $P = 0.0532$. But the interaction is not significant: the data give no evidence of a difference between the periods in the territory sizes of the phenotypes. In contradiction of earlier results (Davis & O'Donald, 1976*b*), no evidence has been obtained to suggest that male phenotypes may differ in their territory sizes. Territory size cannot be a cause of the variation in the males' breeding dates. As we shall see in chapter 10, section 10.4, several independent lines of evidence strongly support the theory that females prefer to mate with melanic males.

Although phenotype does not appear to determine territory size, territory size may still determine breeding date. Davis & O'Donald (1976*b*) found that territory size strongly influenced the breeding dates of experienced males when they formed pairs with new females. Figure 7.7 shows the relationship of territory size to breeding date when the following samples of new pairs are included:

Period	Experience of male	Number in sample
1974–75	New males	31
	Old males	35
1977–1978	New males	14
	Old males	20
	Total	100

From these data, we obtain an overall regression of breeding date on territory size

$$Y = 33.12 - 0.0002623\ x$$

which expresses the mean breeding date Y in terms of territory of size x.

Figure 7.7 Relationship of territory size to breeding date (males). Open squares, pale birds; filled diamonds, intermediates; and filled squares, darks.

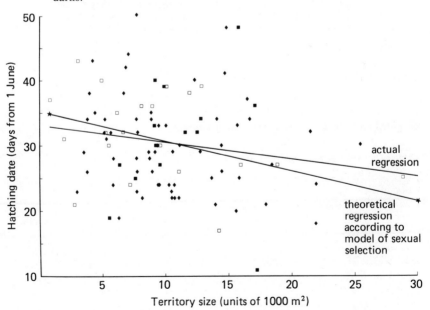

Table 7.5. *Test of homogeneity of regressions in different classes of males*

Source of variation	SSx	SP	SSy	Dfs	Residual SS	Dfs	Regression SS
1974–75							
New males	755 720 524	−260 116	1695.42	30	1605.89	29	89.5310
Old males	739 800 108	−504 151	1608.17	34	1264.61	33	343.563
1977–78							
New males	1 618 530 412	−26 859.9	795.43	13	794.98	12	0.4457
Old males	1 018 036 520	−180 786	657.75	19	625.65	18	32.1044
Total	4 132 087 564	−971 912	4756.77	96	4528.17		
Residuals from individual regressions					4291.13	92	
Difference for homogeneity of regressions					237.04	3	

Test of homogeneity: $F_{3,92} = (237.04/3)/(4291.13/92) = 1.6940$; $P = 0.174$

In this table, SSx is the sum of squares of territory sizes, SP is the sum of products, and SSy is the sum of squares of breeding dates.

Table 7.6. *Test of homogeneity of correlation coefficients in different classes of males*

Period	Experience of male	Correlation coefficient r	Transformed coefficient z_i	Weight $n_i - 3$	Weighted sum of squares $(z_i - \bar{z})^2 (n_i - 3)$
1974–75					
	New males	−0.2298	−0.2340	28	0.1321
	Old males	−0.4622	−0.5001	32	1.2476
1977–78					
	New males	−0.0237	−0.0237	11	0.8562
	Old males	−0.2209	−0.2246	17	0.1035

Weighted mean transformed correlation coefficient:

$$\bar{z} = -0.3027$$

Value of χ^2 for 3 degrees of freedom:

$$\chi_3^2 = 2.339$$

Value of P:

$$P = 0.505$$

This regression equation has been superimposed on the values plotted in figure 7.7. The breeding date is counted in the number of days from 1 June up to the hatching of the first egg. Thus a value $y = 33$ indicates that the pair hatched their first egg on 3 July. The equation shows that a larger territory produces an earlier breeding date. The regression is highly significant:

Source of variation	Sum of squares	Dfs	Mean square	Value of F	Value of P
Regression	292.43	1	292.43	5.6489	0.0194
Residual	5073.28	98	51.768	—	—

This regression analysis gives an estimate of the correlation of breeding date and territory size in new pairs:

$$r = -0.2335$$

Davis & O'Donald (1976*b*), analysing only the data of 1974 and 1975, found a high and significant correlation for experienced males in new pairs:

$$r = -0.4622$$

and a non-significant correlation for new males:

$$r = -0.2298$$

The difference in these correlations is not significant. A test of the homogeneity of regression for samples of both 1974–75 and 1977–78 is given in Table 7.5. The calculation is based on an analysis of covariance in the samples. The residual sums of squares shown in the table are the sums of squares of breeding dates in each sample after the corresponding regression sums of squares have been subtracted. These residuals represent the variation in breeding dates after the effects of territory size have been removed. Steele & Torrie (1980) describe the method in detail.

We can also test the homogeneity of the correlation coefficients shown in table 7.6. This is not the same test as that of the homogeneity of regression coefficients. The correlation is determined by the level of variation in both dependent variate (breeding date) and independent variate (territory size). Only the independent variate determines the regression coefficient. Neither regression coefficients nor correlation coefficients show heterogeneity between samples of males. The test of homogeneity of correlations uses Fisher's z transformation:

$$z = \tanh^{-1}(r)$$

with variance

$$\text{var}(z) = 1/(n-3)$$

where n is the sample size. This transformation has the advantage that z has an approximately normal distribution. The weighted sum

$$\sum_i (z_i - \bar{z})^2 (n_i - 3)$$

is a sum of squares of standard normal variates and hence distributed as χ^2 as shown in table 7.6. This test of homogeneity shows that the higher correlations found between territory size and breeding date when older, experienced males take new mates are not significantly different from the lower correlations found for younger males new to Fair Isle. Both the overall correlation coefficient

$$r = 0.2335$$

and the overall regression

$$Y = 33.12 - 0.0002623x$$

are highly significant, with no evidence to suggest that they vary between the different categories of males in the new pairs. By maintaining a larger territory, both new and experienced males mate earlier in the breeding season. Since earlier pairs fledge more chicks, males with larger territories gain a selective advantage. Sexual selection thus favours males with larger territories.

7.4 Sexual selection and the evolution of territoriality

Whenever a resource, necessary to survival or reproduction, is limited and can only satisfy the needs of a proportion of the individuals in a population, competition must then follow. If individuals vary in their ability to compete for the resource, as inevitably they will, the best competitors will secure most of the resource and exclude the poorest competitors from sharing it. Those characteristics that improve the ability to compete will be selected. This is simply a condensed account of Darwin's explanation of selection (Darwin, 1859). Selection is a deduction from the empirical premises that populations can always increase, that resources are always limited, and that individuals vary in the ways they can exploit particular resources. The conclusion – that some individuals are selected at the expense of others – is also empirical: it can be tested by observation of the different chances of surviving or reproducing. Since the premises are empirical and not entailed by the conclusion, it is astounding that some philosophers (for example, Popper, 1972) should have thought that natural selection is a tautological concept and not empirically testable. Can they ever have read Darwin's works? If so, they can hardly have understood him. Ecological geneticists have repeatedly tested and verified Darwin's theory of natural selection. For example, Darwin's theory predicts that a reduction in resources should increase competition and hence the intensity of selection. The intensity of selection has always been found to depend on the availability of resources. Selection of *Drosophila* for an optimum number of bristles increases in intensity in population cages supplied with a smaller number of food tubes (O'Donald, 1971). This corroborates Darwin's theory. If level of resource in no way determined selection, or if selection decreased as levels of resource were reduced, this would refute the theory. In sexual selection, the limited resource is the limited number of ova that the females can produce. The males must compete to fertilize them. The empirical content of Darwin's theory of sexual selection in monogamous birds is obvious: it depends on ecological and behavioural premises, such as the relationship between breeding date and reproductive success (see chapter 8, section 8.3). Independent lines of evidence can be used to test the theory at each point in its chain of deductive inference. In the remaining chapters of this book, I shall show how data of the Arctic Skuas on Fair Isle corroborate Darwin's theory of sexual selection.

Initially, territoriality may have evolved to secure and defend any of a variety of different resources distributed over space. Territories are often areas in which the occupier forages to feed himself, his mate, or his offspring. Without a territory, a pair of birds would compete with other pairs both near and far from their nest. They would be forced to forage at a greater average distance from their nest than pairs with territories. I have already described (in chapter 3, section 3.4), Andersson & Götmark's (1980) attempt to test this hypothesis by comparing the diets of solitary pairs of Arctic Skuas with the diets of pairs in colonies. All they found was a partial correlation of terrestrial items of diet with solitary nesting. The evolution of territoriality as an adaptation to feeding on a particular diet is not the only explanation of the correlation. The feeding behaviour might equally well have evolved as an adaptation to territoriality. Andersson & Götmark also suggested that the dispersion of nests within territories may protect the chicks against predation. The nidifugous Arctic Skua chicks leave the nest a day or two after hatching. They may be found wandering or crouching anywhere within their territories. Predators seldom take them before they are fledged. Even Bonxies will not penetrate far into a dense colony of Arctic Skuas. But when the young Arctic Skuas are learning to fly, they blunder about on their clumsy first flights. The Bonxies easily catch them if they flop down outside their territories and away from the general protection of the colony. As we saw in chapter 2, section 2.3, the Bonxies on Foula predate a large proportion of the fledgling Arctic Skuas. Territoriality may have evolved initially as a strategy to protect the young against predators. Once it has evolved, it limits the number of males who can hold territories on the breeding grounds. The males must then compete for space in order to mate. Territoriality, by its existence, becomes a mechanism of sexual competition. Sexual selection thus increases the initial selective advantage of defending a territory. Larger territories become sexually more advantageous. Every increase in territory size increases the proportion of males who are excluded from holding territories in the limited space available. The sexual selection becomes more intense. Of course, the territories cannot increase in size indefinitely: they get more and more difficult to defend. Up to a point, the original advantage of a territory – as a place in which to feed or conceal the chicks – is simply reinforced by the consequent sexual selection. But the territories may become too large for their original purpose: then the gain in fitness is offset by the loss in defending an area larger than it need be for feeding or concealment. Eventually the loss in fitness must become so great that it also offsets the additional sexual advantage. No further increase in territory size will then occur: stabilizing selection maintains the mean territory size at an optimum determined by the balance of sexual and natural selection.

This theory can be summarized as follows. Territoriality is a behavioural response to secure a resource that is limited in space. The space may be needed for mating, rearing offspring or feeding. A larger territory provides more of the resource for the occupier, so there is less for others. Larger territories should always be advantageous. But they take more defending. Their defence diverts energy from other activities. A balance is ultimately reached between the gains and losses. When males increase their territory size in a restricted area, a greater proportion of males are left with no territory to defend. They are denied access to the females. Every increase in territory size thus increases the competition for territories. The selection reinforces itself: it acquires a momentum of its own. As we shall see (chapter 8, section 8.2), this is a general characteristic of sexual selection. It can produce a 'runaway process', exaggerating the development of sexually selected characters.

O'Donald (1977) modelled this general process of sexual selection. The model is used to compute the probability distribution of the breeding dates of males with varying sizes of territory. The model is derived from the following premises:

(i)   the males occupy the territories;

(ii)   the females determine the breeding dates;

(iii)   they arrive in succession and mate with one of the remaining unmated males;

(iv) the probabilities of the matings are proportional to the sizes of the males' territories.

The following symbols are used in calculating the probabilities of the matings:

p_{ij} the probability that the jth male is unmated when the ith female arrives ($p_{1j}=1$).

$p_{i+1,j}$ the probability that the jth male is unmated when the $(i+1)$ female arrives.

P_{ij} the probability that the jth male mates with the ith female.

x_j the size of the jth male's territory.

w_j the average number of chicks produced by the jth male.

W_i the average number of chicks produced by the ith female.

y_i the ith female's breeding date.

The probability that the ith female mates with the jth male is proportional to the relative size of his territory, thus

$$P_{ij} = p_{ij}x_j \Big/ \sum_j p_{ij}x_j$$

If $P_{ij} < p_{ij}$

then $p_{i+1,j} = p_{ij} - P_{ij}$

If $P_{ij} \geq p_{ij}$

then $P_{ij} = p_{ij}$ and $p_{i+1,j} = 0$

The breeding success of the males is then

$$w_j = \sum_i P_{ij} W_i$$

The values of W_i are determined by the dates when the females arrive on the breeding grounds. Table 7.7, condensed from table 8.7 in chapter 8, gives the mean numbers of chicks fledged by pairs hatching their eggs in successive weekly intervals in the breeding season. These are the values of W_i. A theoretical fitness function could also be used to determine relative values of W_i. Thus in table 8.7, it is shown that the function

$$W_i = 1 - (\theta - y_i)^2 / \phi$$

fits the data of relative numbers of chicks fledged. In this function the successive intervals of the breeding season, starting with the interval 10–16 June, are symbolized by

$$y_0 = 0$$
$$y_1 = 1$$

Table 7.7. *Breeding date and fledging success of the Arctic Skua*

Date of hatching of first chick	Number of chicks fledged
10–16 June	1.6250
17–23 June	1.5826
24–30 June	1.5128
1–7 July	1.1212
8–15 July	0.7419

and so on. The best fit is obtained when

$$\theta = 0.4735$$
$$\phi = 24.25$$

The generalized fitness function can then be used to investigate how different relationships of fitness to breeding date affect the sexual selection of territories.

To show how the model works, suppose five males with territory sizes 5, 25, 40, 60 and 85 m^2 are to be mated; while five females arrive to breed on days 17, 22, 28, 34 and 39. Table 7.8 shows the bivariate probabilities of mating and the numbers of chicks fledged by the males. The bivariate probability distribution gives rise to a correlation coefficient of -0.632 between territory size and breeding date. The males with the larger territories are much more likely to mate with the earlier females. The male with the very small territory mates with the last female to arrive four times out of five.

The breeding dates and territory sizes of male Arctic Skuas on Fair Isle, previously shown in figure 7.7, can be used to calculate the hypothetical bivariate distribution P_{ij}, hence the hypothetical regression of breeding date on territory size and the hypothetical correlation coefficient.

Davis & O'Donald (1976b), using only the data of old males for the earlier period 1974–75 (a total of 35 values of territory size and breeding date), obtained the theoretical correlation

$$\rho = -0.3941$$

which compared closely with the actual correlation

$$r = 0.4622$$

for these data (see table 7.6). The regression equations, too, are closely similar as shown in the figure in our original paper (and also shown as

Table 7.8. *Example of probabilities of mating and resultant fitnesses when probability of mating is proportional to territory size*

Breeding date of females	Territory size of males (m²)					Chicks fledged by females
	5	25	40	60	85	
17	0.0233	0.1163	0.1860	0.2791	0.3953	1.6250
22	0.0317	0.1433	0.2112	0.2805	0.3333	1.5826
28	0.0488	0.1912	0.2490	0.2729	0.2382	1.5128
34	0.0991	0.3036	0.3129	0.1675	0.0332	1.1212
39	0.7972	0.2457	0.0409	0.0	0.0	0.7419
Chicks fledged by males	0.8643	1.2276	1.3943	1.4981	1.5674	

figure 7.1 in O'Donald, 1980*a*). But we considered that, in spite of this close agreement between the theory and our observations, the behaviour of the females refutes the basic premise of the theory that the probability of mating is proportional to territory size. This premise would hold if females entered territories at random after their arrival on the breeding grounds. Yet, as we have seen (table 7.1 of section 7.1 of this chapter), when females choose a new mate, they often come from the same area where they had previously bred in the colony: they do not spread themselves at random over the colony. Perdeck (1963) observed that females visit different males on their territories to choose their mates. This behaviour is contrary to the premise on which the model is built. The model cannot wholly explain the relationship between breeding date and territory size in the Arctic Skua. Nevertheless, it may explain part of it: many of the females will be new to Fair Isle, breeding for the first time; to the extent that chance determines where they arrive, so they will tend to choose males with larger territories; larger territories will be sexually selected partly in accordance with the theory.

The data of both new and experienced males in both of the periods 1974–75 and 1977–78 consist of 100 observations of breeding date and territory size. Since we found no heterogeneity between males or periods (tables 7.5 and 7.6), all these observations may be combined to test the goodness of fit of the model. The 100 females are assumed to arrive in the order of their breeding dates. They mate with the 100 males on their territories with probabilities proportional to the males' territory sizes. The bivariate probability distribution, P_{ij}, is then calculated for the 100 males and females as in the simple example of table 7.8. P_{ij} is the hypothetical probability distribution, derived from the model, that gives the expected frequencies with which each female mates with each male, hence the hypothetical means, variances and covariance:

$$\varepsilon\,(x) = 10875.1$$
$$\varepsilon\,(y) = 30.2701$$
$$\text{var}\,(x) = 42\ 514\ 672$$
$$\text{var}\,(y) = 53.6769$$
$$\text{cov}\,(x,y) = -19\ 559.5$$

Then we obtain the hypothetical regression equation

$$Y = 35.27 - 0.0004601x$$

and correlation coefficient

$$\rho = -0.4094$$

These can be compared with the observed regression and correlation

$$Y = 33.12 - 0.0002623x$$
$$r = 0.2335$$

The expression

$$t = \frac{b - \beta}{(S_{y\cdot x}^2 / \sum (x - \bar{x})^2)^{\frac{1}{2}}}$$

tests the significance of the difference between the actual and hypothetical regression coefficients, b and β, where

$$S_{y\cdot x}^2 = \text{Residual Mean Square of } y$$
$$= 51.768$$

as found in section 7.3. Hence we find

$$t = -1.7923$$

and for both tails of the distribution $(\pm t)$ this corresponds to

$$P = 0.0762$$

The difference is not significant at $P = 0.05$.

A χ^2 test, based on the same principle as the test shown in table 7.6, shows whether the difference between r and ρ is significant. The transformed correlation

$$z = \tanh^{-1}(r)$$

is an approximately normal variate with expectation

$$\varepsilon\,(z) = \tanh^{-1}(\rho)$$

and variance

$$\text{var}(z) = 1/(n-3).$$

Therefore

$$\chi_1^2 = [z - \varepsilon(z)]^2/\text{var}(z)$$
$$= 3.7685$$

corresponding to

$$P = 0.0522$$

The correlation coefficient is not significantly different from its hypothetical value in the model in which the males' chances of mating are proportional to the sizes of their territories. But the difference is very close to being significant at $P=0.05$.

A random distribution of females over the breeding grounds is not the only possible mechanism by which territory size determines the probability of mating. For example, a male's chances of mating may simply depend on the intensity and persistence of his courtship. This may depend on his level of testosterone, which may also determine his territory size. Territory size would then vary with the chances of mating but would not determine those chances. Sexual selection would raise levels of testosterone and concommitantly increase territory size. The effect would be the same as if territory size determined the chances of mating. The earlier breeding of males with larger territories would be a correlated effect of their more intense and persistent courtship. I previously thought that this correlated effect was produced by the preferential mating of the melanic phenotypes; for, in Davis & O'Donald's data on territory sizes of pairs in 1974 and 1975 (Davis & O'Donald, 1976b), melanic males had significantly larger territories than pale males. But as we saw in section 7.3 of this chapter, the more recent data of 1977 and 1978 contradict our earlier result: the phenotypes show no significant differences when the two sets of data are combined; their overall mean territory sizes are similar. Sexual selection of the phenotypes cannot explain the correlation of territory size and breeding time. This must depend on aspects of sexual behaviour uninfluenced by the melanism. Whatever may be the cause of this correlation, however, the earlier breeding of males with larger territories necessarily produces a selective advantage: territoriality is sexually selected.

8

Sexual selection

8.1 Darwin's theory of sexual selection

Sexual selection occurs when some individuals are more successful in finding mates than others. Like other characteristics of animals, sexual behaviour varies. Some males will defend their territories more fiercely than others, or court the females more vigorously, or display more attractive plumage. These characteristics will be selected if they increase the chances of mating. This is sexual selection as Darwin defined it in *The Origin of Species:*

'This depends, not on a struggle for existence, but on a struggle between the males for possession of the females; the result is not death to the unsuccessful competitor, but few or no offspring. Sexual selection is, therefore, less rigorous than natural selection. Generally, the most vigorous males, those which are best fitted for their places in nature, will leave most progeny. But in many cases, victory depends not on general vigour, but on having special weapons, confined to the male sex. A hornless stag or spurless cock would have a poor chance of leaving offspring'.

And Darwin thought that sexual selection would be the cause of the evolution of many of the differences in structure and coloration of the sexes:

'Thus it is, as I believe, that when the males and females of any animal have the same general habits of life, but differ in structure, colour, or ornament, such differences have been mainly caused by sexual selection; that is, individual males have had, in successive generations, some slight advantage over other males, in their weapons, means of defence, or charms; and have transmitted these advantages to their male offspring'.

Darwin infers that sexual selection has taken place only if the sexes have the same habits of life. If the males and females lived part of their lives in different habitats or engaged in different activities, then natural selection, differing between the sexes, would be sufficient to explain their different structures.

In *The Descent of Man and Selection in Relation to Sex* (1871) Darwin elaborated his theory in great detail, applying it to many examples of sexual differences in insects, fish, birds and mammals. When writing *The Origin of Species*, Darwin was thinking of males competing for females, mainly by fighting for them. This part of his theory has never met serious opposition from biologists: stags obviously do compete; this must produce evolution of strength and weaponry. But some characteristics in which the sexes differ have no obvious role in direct competition. The brilliant plumage of some male birds is displayed only to the females. And Darwin thought the females would prefer to mate with the more highly adorned and hence more attractive males. It is almost always the males that will thus have been selected: they have the weapons; they have the fine and colourful plumage which they display to the females. But why is it only the males? Darwin explained the sexual selection of males in terms that are now fashionable in sociobiological theory:

'The female has to expend much organic matter in the formation of her ova whereas the male expends much force in fierce contests with his rivals, in wandering about in search of the female, in exerting his voice, pouring out odoriferous secretions, etc.: and this expenditure is generally concentrated within a short period . . . . On the whole the expenditure of matter and force by the two sexes is probably nearly equal, though effected in very different ways and at different rates'. (Darwin, *The Descent of Man*, chapter 8).

The male, being the producer of the microgametes, has an enormous reproductive potential which is limited by the relatively few macrogametes of the female. This gives rise to sexual selection. The males must compete: the females can choose.

But female choice – the expression of female preference in favour of male adornment and male display – was strenuously denied as a mechanism of sexual selection. Darwin himself could only provide anecdotal evidence that certain females would indeed only mate with particular males and not with others – individual female pigeons showing strong preferences for males of particular breeds or colours, for example. In terms of the functional significance of male plumage, Darwin's general argument for female choice is powerful:

'Now with birds the evidence stands thus: they have acute powers of observation, and they seem to have some taste for the beautiful both in colour and sound. It is certain that the females occasionally exhibit, from unknown causes, the strongest antipathies and preferences for particular males. When the sexes differ in colour or in other ornaments the males with rare exceptions are the more decorated, either permanently or temporarily during the breeding season. They sedulously display their various ornaments, exert their voices, and perform strange antics in the presence of the females. Even well-armed males, who, it might be thought, would altogether depend on the law of battle, are in most cases highly ornamented; and their ornaments have been acquired at the expense of some loss of power. In other cases ornaments have been acquired, at the cost of increased risk from birds and beasts of prey. With various species many individuals of both sexes congregate at the same spot, and their courtship is a prolonged affair. There is even reason to suspect that the males and females within the same district do not always succeed in pleasing each other and pairing.

What then are we to conclude from these facts and considerations? Does the male parade his charms with so much pomp and rivalry for no purpose? Are we not justified in believing that the female exerts a choice, and that she receives the addresses of the male who pleases her most? It is not probable that she consciously deliberates; but she is most excited or attracted by the most beautiful, or melodious, or gallant males'.

But biologists were reluctant to accept that females could choose. And it was precisely that the male characteristics were beautiful – like the tail of the peacock, the primary wing feathers of the Argus pheasant, or the plumes of the bird of paradise – that was the difficulty. We can see these things as beautiful. Can a female bird have so conscious an appreciation of beauty, similar to our own, that she will choose to mate with the most beautiful males, so that sexual selection will enhance their beauty in succeeding generations? Darwin foresaw this would be raised as a fatal objection to his theory. He tried to forestall it by his argument, which he elaborated further, that the females' choice would not depend on a conscious appreciation of beauty, but rather on the excitement – the increased sexual stimulation – produced by brilliant displays of brightly coloured plumage. Such an explanation can easily be stated in behavioural terms. Female mating response is elicited at a certain threshold level of stimulation provided by male courtship. Brilliant displays stimulate more than dull ones; the threshold is more rapidly reached; hence females prefer to mate with the more brilliant males. Females would exercise choices between different types of male plumage if they responded to one plumage type at a different threshold to another. Some females would choose one type of plumage, others another type, in the males they mated with.

In two influential papers, Huxley (1938a, b) confused the discussion of sexual selection that had only recently been clarified by Fisher's *The Genetical Theory of Natural Selection* (1930). Fisher had sought to explain just why the females should mate preferentially at all. He put forward the evolutionary theory of the origin of mating preference that I shall discuss in the following section. Huxley quoted Fisher, but misunderstood his theory, denying the possibility of sexual selection by female choice. The confusion in Huxley's thoughts on the subject, I have already discussed (O'Donald, 1980a). And in view of the general acceptance of female choice in mate selection, Huxley's two papers are now only of historical interest. Perhaps now, indeed, fashion, having swung back towards Darwin's view, has gone well beyond it. Sociobiologists often seem prepared to assume that any behavioural character, if it can be shown to give rise to some selective advantage, will then necessarily have a genetic basis, and hence have evolved.

The assumption is often made that all behaviour will be selected to its optimum level of expression. And, as a corollary, behaviour that seems optimally expressed must therefore have evolved by natural or sexual selection. This assumption takes for granted that genes can do anything, so that any character can evolve. For example, it has been suggested that females should assess a male's worth by the quality of the territory he is defending. Females would thus gain an advantage by choosing to mate with the males with the best territories: hence they will evolve preferences determined by territorial quality. The question is asked: do females more often choose the males with the better territories? To answer it, territorial quality must be measured. Quadrats are put down in a suitable random pattern. Food plants are sampled. Calorific values are measured in bomb calorimeters. Elaborate statistical methods may be used to combine the values from the different food plants allowing for associations in the quadrats. It may be found that the males' territories do indeed differ in the quality and quantity of the food they can provide. But although the ecologist has had to use advanced modern technology to establish this fact, yet he may conclude that males do try to hold high quality territories and females do choose to mate with them for this reason.

How far can adaptation really be taken? How can genes determine an instinct to discern what cannot be revealed without the application of modern technology? This is not like the evolution of complex sense organs to detect a direct stimulus. It is the evolution of behavioural responses to environmental differences only detectable by the use of a range of special methods. To suppose that Darwinian evolution can do virtually anything is indeed the caricature of Darwinism that Lewontin (1977) attacked in his review of Dawkins' *The Selfish Gene* (1976). As an attack on Dawkins' book, this was grossly unfair, for Dawkins nowhere postulates the evolution of behaviour that an animal is unlikely to be able to perform. Lewontin's attack was apparently ideological, based on the Marxist assumption that behaviour is largely determined by the environment. Dawkins certainly assumes that behavioural differences are genetically determined – like the differences between the sexes in sexual behaviour. But this is the necessary starting point in any discussion of evolution.

Unrealistic and too simple though it may be, one must start somewhere. Genetic variation is, after all, the prior condition for the possibility of evolutionary change.

We must assume, therefore, that mating preferences for particular phenotypes can be genetically determined. We know from many experiments, especially with insects and fish, that mating preferences are expressed for particular characteristics. As humans who depend so much on sight, we can most easily appreciate those characteristics that are visibly expressed; often they are the alternative forms in a genetic polymorphism. Other organisms depending on other senses will more easily appreciate non-visual characteristics, less obvious to us. If some females can discriminate between the males by the characteristics they display, their ability to discriminate, coupled with a propensity to mate with a particular phenotype of male, might have a genetic basis. If so, discrimination and mating propensity could be selected. This is the starting point of Fisher's theory of the evolution of mating preferences.

8.2 Sexual selection and the evolution of mating preferences

Fisher's theory of the evolution of mating preferences was the first new development in the theory of sexual selection since Darwin's work.

The whole of Fisher's theory is stated in four paragraphs of *The Genetical Theory of Natural Selection* (Fisher, 1930), chapter 6. Fisher's statement of his theory is concise enough to be quoted fully:

'Whenever appreciable differences exist in a species, which are in fact correlated with selective advantage, there will be a tendency to select also those individuals of the opposite sex which most clearly discriminate the difference to be observed, and which most decidedly prefer the more advantageous type . . .

'Certain especially remarkable consequences do follow if some sexual preferences of this kind, determined, for example, by a plumage character, are developed in a species in which the preferences of one sex, in particular the female, have a great influence on the number of offspring left by individual males. In such cases the modification of the plumage character in the cock proceeds under two selective influences (i) an initial advantage not due to sexual preference, which advantage may be quite inconsiderable in magnitude, and (ii) an additional advantage conferred by female preference, which will be proportional to the intensity of this preference . . . .

'The two characteristics affected by such a process, namely plumage development in the male, and sexual preference for such developments in the female, must thus advance together, and so long as the process is unchecked by severe counterselection, will advance with ever-increasing speed. In the total absence of such checks, it is easy to see that the speed of development will be proportional to the development already attained, which will therefore increase with time exponentially, or in geometric progression . . . .

'If carried far enough, it is evident that sufficiently severe counterselection in favour of less ornamented males will be encountered to balance the advantage of sexual preference; at this point both plumage elaboration and the increase in female preference will be brought to a standstill, and a condition of relative stability will be attained'.

Now why does Fisher think that this 'runaway process' of sexual selection should continue to produce further development in the ornamented plumage character of the males until finally stopped by counterselection? Fisher has shown, though without genetical argument, that the evolution of the female preference will add to the selective advantage of the preferred plumage characteristic: selection for the preferred plumage will get stronger and stronger. But he has not shown why more extreme developments should automatically be selected. The evolution will go more quickly. Why should it go further? The preference originally evolved for the character that the females could discriminate. This was a character indicating higher fitness in the males. Perhaps linkage or a pleiotropic effect may determine the appearance of the character in the fitter males: it gives the females the possibility of discrimination. Fisher's argument shows that if some females do discriminate and mate with the fitter males, they gain an advantage, increase in frequency, and thus add a proportionate increase to the fitness of the males. I cannot see why further developments in plumage should also be preferred unless the females' preference is relative. If the females respond to more extreme developments as to a supernormal stimulus, then new genes, producing increased development would be preferred and selected. Arguments can be put forward to support this idea (O'Donald, 1980a) assuming that a shift in female response will evolve as part of a general behavioural adaptation to favourable stimuli. The position of maximum response to stimuli may, after training by reward and punishment, be shifted to a point above the level at which the favourable stimulus was originally rewarded during training. This may be a general adaptation of behavioural response, if the penalty for not responding to favourable stimuli at a particular level or intensity is generally greater than the penalty for responding to stimuli at higher levels or intensities. If more extreme sexual characters act as stimuli or supernormal stimuli in eliciting the sexual responses of the females, the females, having evolved preferences for a particular level of development of the male character, will also prefer more extreme developments. New alleles, increasing the development of the character, will spread through the population, gaining an advantage from the same preference that selected the lesser developments of the character. Fisher's runaway process will continue for as long as an overall advantage is gained and natural selection is not too adverse.

Fisher's argument for runaway sexual selection is valid only if female preferences are relative. Even if this is granted, the argument is still not rigorous: it must be stated in genetical terms in order to demonstrate that the premises do entail the conclusions. Fisher's own statement of the theory does not incorporate a genetical model. Darwin himself could have put forward the theory in the terms Fisher used. It is easy to see where difficulties may lie. Fisher assumes that the males are selected by the females: the females' preferences are passed to their sons as well as daughters. The male offspring of the preferential matings possess the preferred character. The preference expressed for them in the next generation selects also the preference they inherited from their mothers. But the rate of selection of the preference is exponential, as Fisher suggests, only if it is carried by the same proportion of the preferred males and the preference increases in a constant ratio together with the preferred males. Genetically, this cannot happen. The increase in the ratio of preferred to non-preferred males may be exponential at the start, but does not remain so after the preferred males have become common. By genetic recombination, the genes for the females' preference will be passed to the disfavoured males as well as the others. We may expect some association to develop between the genes for the preferred character and the preference. But this will be determined by the relative frequencies of the genes as well as the preferential mating rates. Genetic models show that the association between the genes usually increases at first while the genes are rare, but declines later when they have become common. There is no constant exponential rate of increase or constant genetic association of preference and preferred character that would lead to a geometric increase in the rate of selection.

Genetic models of Fisher's theory are complicated to construct and cannot be analysed in detail without extensive computer simulations. But some general conclusions can be obtained by comparing the numerical results of selection in different models. I have already described some models and detailed results in chapter 8 of my book on sexual selection (O'Donald, 1980*a*). Rates of selection are usually slower, sometimes much slower, than Fisher suggested: the rate of selection of preference is geometric only at the start of the process when the genes for the preference are rare in the population. Even then, selection will still be very slow if a dominant gene determines the preferred character of the males. Only when females prefer a recessive do rates of selection increase rapidly: only then is the preference selected to high frequencies in the females.

At the start of selection, as Fisher argued, natural selection is necessary, or the process may not get going at all. Once the females with preferences have increased in frequency, natural selection may add little to the rate of increase produced by the sexual selection. If preferences are relative, then a new gene increasing the attractiveness of the males will also be selected: as it spreads through the population, the preference also increases, usually doubling or trebling its frequency in the course of the selection of a gene for the preferred character. For the preference to evolve to high frequency, several stages may be necessary, at each of which a new gene, adding to the attractiveness of the males, replaces a previously selected gene.

In general we might expect that each new gene would evolve dominance in the course of its evolution to become a new wild-type. But in sexual selection, this is not to be expected. In the first place, computer simulation shows that a recessive character would be selected more rapidly than a dominant (O'Donald, 1980a). Recessives would tend, therefore, to be picked out by sexual selection. In the same way, modifiers producing recessiveness might also be selected, just as modifiers producing dominance will be selected in the spread of a gene by natural selection.

Finally, we should expect that some degree of assortative mating will be an outcome of sexual selection. Since the preference evolves because it becomes associated with the preferred character, the females who possess the preferred character will tend to be those who express the preference. The preferred character will therefore show assortative mating. Many sexually selected characters are of course sex-limited to males. Assortative mating will not be detectable unless the genotypes for the male characteristics can be detected in the females by biochemical or other tests.

Fisher's general theory certainly explains the persistence of specific female preferences.

With the additional premise that the more extreme developments in the preferred character will act as supernormal stimuli to the females, the theory then explains the evolution of those extreme and exaggerated characters for male display such as the peacock's tail or the Argus pheasant's wing feathers. The chain of reasoning is thus complete. Darwin was unable to explain the origin and persistence of specific preferences. He had no evidence or argument to confute the objection that the females' preferences would be fickle. He knew that females really did make choices, but he could only assert that their choices would be consistently maintained. To explain the expression of preference for ever more extreme developments of a male character, we have been obliged to assume that these act as supernormal stimuli to the females. But we know that supernormal stimuli do produce exaggerated responses: this is an empirical fact, though its cause in the evolution of behaviour may not yet have been fully explained. For our theory, it remains a necessary premise if the more extreme forms of sexual dimorphism are to be explained by sexual selection.

Many polymorphic birds, like the Arctic Skua, are not sexually dimorphic. Both sexes develop the different phenotypes of plumage. If brightly coloured polymorphic characters are shown only in the males – for example the polymorphisms of male guppies or ruffs – then sexual selection may be presumed to take place. But sexual selection may still take place if both sexes are polymorphic. Assortment in the matings of polymorphic forms is, prima facie, evidence for preferential mating. Assortative mating is the tendency of the same phenotypes to mate with each other. It is the mating of like with like. It cannot be explained plausibly by male competition. Males are so often indiscriminating when offered a choice of females to mate with that it seems highly unlikely they would compete more fiercely, just to mate with females like themselves. This would make no sense in terms of a given investment in time and energy required to produce as many offspring as possible. Female choice is the most likely explanation of assortative mating, especially since any genetically determined propensity to preferential mating will evolve towards an assortative expression of preference. The expression of the character in both sexes is not an argument against sexual selection. We should certainly expect that extreme and exaggerated characters would have evolved so as to be expressed only in males. These characters must increase the chances of predation: the male peacock is more easily seen; his cumbersome tail must hinder his escape. Red-throated male sticklebacks are much more often taken by predators than the sexually disadvantaged non-red males (Moodie, 1972; Semler, 1971). It is this increasing risk of predation that must ultimately bring the runaway process of sexual selection to a halt: at some point the mating advantage gained by having an even more extreme or elaborate display is balanced by the disadvantage of the increased mortality.

But where characters have not been developed to the point where they increase mortality, no advantage is gained by limiting or reducing their expression in the females. Selection for male sex-limitation will only take place when the females, who stand to gain no advantage in mating, start to suffer increased mortality by sharing the same characters with the males. Sexual selection in the males will produce corresponding evolution in the females until increased mortality in both sexes selects modifiers that limit further evolutionary development to the males. Because the selection of modifying genes is always a slow process – it depends on the rate of selection of the character undergoing modification – it will necessarily lag a long way behind the sexual selection itself. The females will inevitably suffer a disadvantage until all those evolutionary developments that increase the risk of predation – or mortality generally – have evolved sex-limited expression in males alone.

Arctic Skuas suffer some disadvantage in having a dark or intermediate phenotype. As we have seen (chapter 6) the dark and intermediate birds breed for the first time in their fourth or later years; pales may breed in their third year, gaining an advantage from their greater chance of surviving to breed. Natural selection in favour of pales thus opposes the sexual selection in favour of darks and intermediates. But there is no evidence or likelihood of increased predation of dark birds. Their later onset of maturity must be a pleiotropic and physiological effect of the genes for plumage. The condition of being dark, intermediate or pale probably has very little effect on survival. No advantage will be gained by limiting to males the sexually selected plumage preferred by the females. But modifying genes will presumably be selected in both sexes to raise the physiological fitness of each plumage genotype. In the Arctic Skua, sexual selection may have been stopped by adverse natural selection. If so, this adverse natural selection is acting not on the sexually selected character itself, but on the pleiotropic effects of the genes that determine it: sexual selection has not been allowed to produce an extreme or exaggerated character that reduces the chances of survival. This is implied by the absence of sexual dimorphism. If the polymorphism in plumage is maintained by sexual selection, there must either be a balance of selective forces acting on genes with pleiotropic effects on different characters – plumage on the one hand, physiological processes of survival and maturity on the other – or different plumage characters must be objects of female preference.

8.3 Sexual selection in polygynous and monogamous birds

Examples of extreme sexual dimorphism are usually to be found in polygynous species. Sexually dimorphic species of game birds are highly polygynous. Even more striking are some examples of dimorphic and polygynous species in groups of mainly monomorphic and monogamous species. Among waders, the ruff stands out from the rest as a highly dimorphic and polygynous species. The male ruffs are polymorphic for the plumage colours of their ruff and ear-tufts. They display these characters vigorously to the females. This must surely be the result of sexual selection. Experimental studies on fish and insects provide the best direct evidence for sexual selection, for choices can be offered and frequencies of matings observed. The male guppy (*Poecilia reticulata*) is a brilliantly coloured little fish. It is highly polymorphic, though the duller coloured females are not. It is also highly polygynous. The males devote a great deal of their time to sexual display and courtship. Matings show the 'rare-male' effect: females tend to mate more often with a particular male phenotype when it is rare than when it is common. This effect is to be expected when females exercise mating preferences. If we assume that certain proportions of females have preferences for different male phenotypes, a rare phenotype of male has relatively more females preferring it than a common phenotype of male. The rare male should mate more often than the common one, simply because relatively more females will be looking for him. Guppies show exactly this effect in the relative frequencies of the males' matings (Farr, 1980). Intensity of male courtship also has a strong effect on the chances of mating.

The rare-male effect was first noted in *Drosophila* by Petit (1954) and has been intensively studied by Ehrman and Spiess for many specific genetic factors in different species of *Drosophila* (Ehrman, 1968, 1972; Spiess 1968; Spiess & Spiess, 1969). It is this effect that produces the stability of many polymorphisms maintained by sexual selection. It is a product of female preference combined with polygyny: in effect, the males are sampled 'with replacement' by the females. The preferred ones may be sampled many times, especially if the females are very reluctant to mate except with those males they prefer. This process leads to considerable variation in male reproductive success: usually the variation in male fitness is very much greater than variation in female fitness.

Higher levels of polygyny produce greater potential variation in male fitness. Males with a harem may mate many times more often than those without, many of whom may not mate at all. This difference in reproductive success may not be as extreme as it might appear, however. Darwin (1871) knew that females in the harem would often have the opportunity, whilst one stag was defending his harem, of escaping with some other male, or at least pairing temporarily with him. Clutton-Brock, Albon, Gibson & Guinness (1979) have documented the proportions of matings that can be achieved by the different stags. Clearly the greater the polygyny, the more intense sexual selection can be. The polygyny of a species can be measured by the average sex-ratio of individuals taking part in matings. Clutton-Brock, Harvey & Rudder (1977) found that in primates sexual dimorphism in size was related to this measure of the 'socionomic' sex ratio: in polygynous primates, males are often twice the size of the females; in monogamous primates they may only be slightly larger.

It is easy to appreciate the sexual advantage that some males might gain as a result of polygynous matings. But many birds are monogamous during a single breeding season. Unless there is a disparity in the adult sex ratio, the males' reproductive success must vary exactly as the females'. Sexual selection can only take place if some males consistently mate with the more fertile or successful females, while others mate with the less fertile or successful. Although the overall variation in reproductive success may be exactly the same in both males and females, some male phenotypes may have a higher mean success rate than others. Darwin put forward a subtle theory, based on a premise of the breeding ecology of birds, to explain how different male phenotypes might vary in their mean reproductive rates. In Darwin's words, the theory is as follows:

'Let us take any species, a bird for instance, and divide the females inhabiting a district into two equal bodies, the one consisting of the more vigorous and better-nourished individuals, and the other of the less vigorous and healthy. The former, there can be little doubt, would be ready to breed in the spring before the others; and this is the opinion of Mr. Jenner Weir, who has carefully attended to the habits of birds during many years. There can also be no doubt that the most vigorous, best-nourished and earliest breeders would on an average succeed in rearing the largest number of fine offspring. The males, as we have seen, are generally ready to breed before the females; the strongest, and with some species the best armed of the males, drive away the weaker; and the former would then unite with the more vigorous and better nourished females, because they are the first to breed. Such vigorous pairs would surely rear a larger number of offspring than the retarded females, which would be compelled to unite with the conquered and less powerful males, supposing the sexes to be numerically equal; and this is all that is wanted to add, in the course of successive generations, to the size, strength and courage of the males, or to improve their weapons'.

Darwin also explained that the advantage of early breeding would produce sexual selection by female preference; for

'. . . the more vigorous females, which are the first to breed, will have the choice of many males; and though they may not always select the strongest or best armed, they will select those which are vigorous and well armed, and in other respects the most attractive. Both sexes, therefore, of such early pairs would as above explained, have an advantage over others in rearing offspring; and this apparently has sufficed during a long course of generations to add not only to the strength and fighting powers of the males, but likewise to their various ornaments or other attractions'.

The correlation of earlier breeding date with increased reproductive success that Darwin postulated is now clearly established for many birds. I have used the empirical relationship of breeding date to breeding success in the Arctic Skua to set up general models of sexual selection in a computer

(O'Donald, 1972*b*, 1973, 1974). In these models, the breeding season is divided into intervals of weeks or days. The breeding success of a pair that mates in a certain interval is given by the average number of chicks fledged by all the pairs in the same interval. The matings take place according to the choices of the females. Some females may have preferences for particular phenotypes among the males. The females who come into breeding condition in a particular interval choose from among all the males who have not yet found mates. If they have preferences, they choose to mate with the phenotypes they prefer. But if none of the preferred males are left unmated, they mate at random with any of the other males. Of course, the earlier females have a wider choice of males: they are more likely to find males they prefer. Towards the end of the breeding season, however, the preferred males have probably all been chosen by the earlier females: only disfavoured males are left unmated. On average, the preferred males mate sooner than the others. They gain a selective advantage by the greater number of chicks they produce as the consequence of mating earlier in the season. Their average selective advantage may be quite large. It is also frequency-dependent. When only a few preferred males are present in the population, they all quickly get chosen by the females that prefer them. They may all be mated right at the beginning of the breeding season. All the other males breed through the rest of the season, with about the overall average reproductive success. Thus the preferred males will gain an advantage if the greatest reproductive success is gained by the earliest breeders. As the preferred males increase in frequency, their relative advantage will change. If most of the females mate preferentially whenever they can, the preferred males gain an advantage that increases as they increase in frequency. This is because the preferred males always get chosen before the others. When the other males have become rare, they get left unmated until the very end of the breeding season. Since the end-of-season matings are often very unsuccessful the disfavoured males, mating last, produce a much smaller than average number of chicks and suffer a large selective disadvantage. Thus when most of the females have preferences for a particular phenotype, the selection is positively frequency-dependent, the selective advantage of the preferred males increasing with their frequency. But if only a small proportion of females have preferences, the opposite effect is shown. As they become common, the number of preferred males greatly exceeds the number of females that prefer them. Some of them get left unmated until the end of the season like the disfavoured males. Then they tend to lose the selective advantage they started with when they were rare: the selection is negatively frequency-dependent.

In the case when only a small proportion of females have any preference (when the proportion is less than 30%), selection can act to favour the preferred males even when early breeding is as disadvantageous as late breeding. The effect of female preference is to produce a distribution of breeding dates of disfavoured males that is more strongly skewed towards the end of the season than the distribution of preferred males is skewed towards the beginning. The preferred males thus gain a slight advantage. Their advantage is much less than in the case when reproductive success declines consistently from the beginning to the end of the breeding season. Seabirds like the Arctic Skua generally do show a consistent decline of reproductive success through their breeding seasons. But passerines may produce relatively unsuccessful clutches early in the breeding season. Sexual selection could still occur, though it would probably be very weak.

These results were all obtained by detailed computer simulations of the sequences of the different matings through the successive intervals of the breeding season (see O'Donald 1972*b*, 1973, 1974, for details of the models).

The final result of sexual selection is the same in both polygynous and monogamous species. If females prefer different male phenotypes, a polymorphism will be established: the phenotypes will remain in the population at stable frequencies to which they will return if moved away from the point of equilibrium. The two mating systems do, of course, produce different rates of sexual selection: the approach to equilibrium is much more rapid with polygyny than with monogamy. A general theory can be developed to determine the results of sexual selection in monogamous birds. This theory is described in section 9.2 of the next chapter. The stability of any polymorphism that may exist depends on the mechanism of mate selection and behaviour. As we shall see (chapter 10, section 10.4) only certain mechanisms of sexual selection are consistent with the maintenance of a stable polymorphism in the Arctic Skua.

8.4 Sexual selection in the Arctic Skua

So far I have described the general theory of sexual selection. Does this theory apply to the Arctic Skua? Are the dark, intermediate and pale phenotypes selected in accordance with the general qualitative predictions of the sexual selection theory? This is the question I asked when I began to investigate the polymorphism of melanic and non-melanic Arctic Skuas. Do some phenotypes have a better chance of mating and obtain mates sooner than others, and do they then gain selective advantage, as Darwin suggested, from the greater reproductive success of breeding earlier in the season? If sexual selection does occur, then we can ask supplementary questions about the behavioural mechanisms of sexual selection: I leave these supplementaries to be answered after genetic models of sexual selection have been analysed in chapter 9. These models are based on specific mechanisms of mate selection which determine the frequencies of matings and the outcome of selection. In chapter 10, I ask: how well do the breeding data on the Arctic Skuas' chances of mating fit the specific models? This leads in sections 10.2 and 10.3 to the estimation of the parameters of the models and tests of their goodness of fit.

To prove that males are sexually selected, it must be demonstrated that

(i) some male phenotypes find mates sooner than others and thus gain a selective advantage by breeding earlier in the season;

(ii) the differences in breeding dates between phenotypes are found only in males, not in females; and

(iii) the differences in breeding dates arise from differences in the chances of mating and are therefore found only when a male is taking a new mate, not when he mates with the same female as in previous years. In the analyses given in the following sections, all the data have come from the Arctic Skua study on Fair Isle.

8.4.1 Males pairing with a new female

Do the males in new pairs have different breeding dates? To answer this, breeding dates must be measured. As I described in chapter 1, the data are recorded on cards for each bird showing its matings in each year by the nest number of the pair. The nest number of a pair identifies its nest record card. This gives details of clutch, hatching dates of eggs, ringing and fledging of the chicks. Sometimes laying date can be ascertained. A nest with a single egg may be found, and a second egg found a day or two later. But pairs can be identified for certain only when their nests have been located. Often they do not attack or show distraction behaviour until their first egg has been laid.

Only when eggs have been laid can traps be used to catch the adults on the nest. Most pairs are not found until they have two eggs. The clearest breeding date that can be defined for almost all pairs is the date when an egg hatches. Since some pairs lay only one egg, breeding date has been measured by the date of hatching of the first egg.

Table 8.1 shows the distributions of breeding dates of males in new pairs when the breeding season is divided into six weekly intervals. Pales are very

Table 8.1. *Breeding dates of melanic and non-melanic male Arctic Skuas in new pairs*

Breeding dates in weekly intervals:	Numbers of males breeding in each interval						
	Period 1948–62			Period 1973–79			
	D	I	P	D	I	P	Total
10–16 June	1	2	1	1	2	0	7
17–23 June	8	24	3	7	21	1	64
24–30 June	16	19	6	14	48	10	113
1–7 July	10	24	6	7	24	10	81
8–14 July	3	15	5	2	10	10	45
15–21 July	1	2	5	2	8	0	18
Total of phenotypes	39	86	26	33	113	31	328
Total in each period		151			177		

D represents dark males, I intermediate males, and P pale males.

significantly later than the others. Measured in the number of days from 1 June, the mean breeding dates are as follows:

Melanics (dark and intermediate)

$\bar{x}_M = 29.44$

Non-melanics (pale)

$\bar{x}_P = 33.33$

The non-melanics take about four days longer to find a mate than the melanics. The mean breeding dates of each phase of male and an analysis of variance of the means and individual values are shown in table 8.2. The mean squares represent estimates of variance caused by differences that may exist between particular effects – the effects of differences between periods and between phenotypes – in addition to the residual mean square of individual values. The residual mean square is calculated from the squares of individual deviations from the means of each phenotype and period. Differences between phenotypes and periods have not contributed to this mean square. If different phenotypes, for example, do give rise to differences in breeding date, the mean square corresponding to differences between phenotypes will be greater than the residual mean square. If not, the F ratio of the two mean squares should have an average value of 1.0. We see that in this case $F = 7.37$. On the 'null hypothesis' that we expect $F = 1.0$,

Table 8.2. *Breeding dates of males in new pairs in periods 1948–62 and 1973–79*

(i) Mean breeding dates

	Phenotype			
	Dark	Intermediate	Pale	All phenotypes
Period 1948–62	27.97	29.93	33.92	30.11
Period 1973–79	28.76	29.77	33.03	30.15

(ii) Analysis of variance of breeding dates

Source of variation	Sum of squares	Dfs	Mean square	Value of F	Value of P
Between periods	0.4878	1	0.4878	0.00825	0.928
Between phenotypes	871.4588	2	435.73	7.370	7.42×10^{-4}
Interaction	22.9296	2	11.465	0.1939	0.824
Residual	19 037.4480	322	59.123	—	—

In this and subsequent tables of analysis of variance, 'Dfs' refers to degrees of freedom. The interaction mean square measures the variance that arises because the phenotypes have different mean breeding dates in the two periods when the Arctic Skuas were studied.

we calculate the probability that a value $F = 7.37$ could arise by the accidents of random sampling. Since this probability is $P = 0.000724$, we can very confidently reject the null hypothesis: the phenotypes differ in their breeding dates at a high level of significance, since it is very improbable that the null hypothesis could be true. But the data do not reject the hypothesis that the breeding dates differ between the two periods. The phenotypes have almost exactly the same mean breeding dates in both periods. Furthermore, the differences between the phenotypes are roughly the same in both periods; no significant interaction is found in the analysis of variance.

In this analysis, and all other analyses in which the same phenotypes are classified into different groups, the interaction in the two-way classification gives rise to a sum of squares that cannot be obtained simply by subtraction as in the standard formulae for the analysis of variance. The differences between phenotypes are not independent of the difference between periods because the numbers in the classes are unequal. A more complicated procedure is necessary to obtain an orthogonal analysis with additive sums of squares. Appendix B gives the details of the computational procedure that has been used in this and the other analyses of variance in this book. In a previously published paper (O'Donald, 1980c) an approximate method

Table 8.3. *Breeding dates of males in new and second-year pairs*

(i) Mean breeding dates of the males

Status	Dark	Intermediate	Pale	All phenotypes
New pairs	28.33	29.84	33.44	30.13
Second-year pairs	21.04	22.55	21.04	22.00

(ii) Analysis of variance in breeding dates

Source of variation	Sum of squares	Dfs	Mean square	Value of F	Value of P
Between new and second-year pairs	5674.0978	1	5674.10	112.1	3×10^{-10}
Between phenotypes	244.1694	2	122.08	2.412	0.0907
Interaction	354.3660	2	177.18	3.501	0.0310
Residual	23 280.6271	460			

of analysis was used, and the sums of squares for phenotypes and interactions are somewhat in error. There are thus small discrepancies between the analyses in this book and in the paper. The source of this error is explained in the appendix.

8.4.2 Comparison of males in new and second-year pairs

The data of table 8.2 show that males in new pairs differ in the average dates they breed. The test of sexual selection is given in table 8.3. This table compares the males' breeding dates in new pairs and in pairs in which they have stayed with the female whom they mated with in the previous year. Only the new and second-year pairs can be used in this test, for third-year and older pairs always breed within a day or two of their breeding date in previous years and thus give no additional independent data on breeding date.

In table 8.3, only the data of second-year pairs in the period 1973–79 have been used to compare the breeding dates. The data of second-year pairs are not consistent between the two periods 1948–62 and 1973–79. Second-year pairs in 1948–62 seem to have bred somewhat later than in 1973–79, although there is no difference at all in new pairs. This is probably explained by the inclusion of some new pairs in the 1948–62 data of second-year pairs. In the earlier period, adults were not all immediately colour-ringed when they entered the colony. They were not all positively identified as belonging to the same pair in a subsequent year. Loss of colour rings also occurred frequently in the early period. Some old pairs were presumed to be old on the basis of their territory and phenotypes.

Undoubtedly some of the presumptive second-year pairs were really new pairs. In the later period, more effective methods of trapping made possible the catching of almost all new birds in the first year they entered the colony. They were colour-ringed with tough plastic rings glued with powerful adhesive. These new colour rings are very durable and very seldom lost from one year to the next. Pairs staying together were easily identified, without the need to make any subjective judgments. The mean breeding dates of second-year pairs are as follows:

1948–62   1973–79
24.48   22.00

The difference in these means is highly significant. The F test gives the probability

$P = 0.000148$

This problem of a possible bias does not arise in the identification of new pairs. No pair was included in the data of new pairs if it might have been a pair from a previous year. Pairs with new territories or with different phenotypes must certainly be new to the colony. The breeding dates of new pairs are almost the same in both the periods: the data of new pairs are completely homogeneous.

To show that the differences in the males' breeding dates are due to sexual selection I therefore compared the mean dates in all new pairs with the mean dates in second-year pairs of the 1973–79 period. This drastically reduces the number of breeding dates of second-year pairs used in the comparison. There were 110 values obtained in the period 1948–62 which have thus been excluded from the data, leaving the 138 values for 1973–79. It is from these values that the means of second-year males shown in table 8.3 have been calculated. Figure 8.1 compares the distributions in new and second-year pairs. The analysis of variance of the data shown in the table then tests the statistical significance of the differences. Of course, the mean breeding date of all phenotypes is completely different in new and second-year pairs. We have already shown that in new pairs the males' breeding dates differ very significantly between the phenotypes. But they do not differ significantly in second-year pairs. Using the data of second-year pairs alone, I obtain the following analysis of variance:

Source of variation	Sum of squares	Dfs	Mean square	Value of F	Value of P
Between phenotypes	72.262	2	36.131	1.156	0.318
Residual	4219.738	135	31.257	—	—

This test, showing no significance in the males' breeding dates, is not sufficient to prove that only in new pairs do the males show different breeding dates. The crucial test of sexual selection is the interaction between the breeding dates in new and second-year pairs and the breeding dates of the phenotypes. Table 8.3 shows that this interaction is significant with probability $P = 0.03$. If we subtract the mean breeding dates of the phenotypes in second-year pairs from their mean breeding dates in new pairs, we thus obtain, for each phenotype, the mean difference of its breeding date in new and second-year pairs. The interaction mean square measures the variance of these differences between phenotypes. Since sexual selection can only operate when pairs are being formed, only new pairs should show a phenotypic variation in breeding date. The significant interaction confirms this conclusion.

In second-year and older pairs, both birds return to their previous territory. They usually arrive back within a few days of each other. The first bird to arrive is often the male. He waits on the territory for the female. If she fails to arrive, he starts looking for a new mate. He associates with females on nearby territories, sometimes bringing about a divorce between another pair and sometimes mating with an unattached female. This takes time: the later breeding dates of new pairs reflects the time a male takes to court and mate with a new female. Second-year pairs, as we have seen,

Figure 8.1 Distributions of breeding dates of new and second-year pairs.

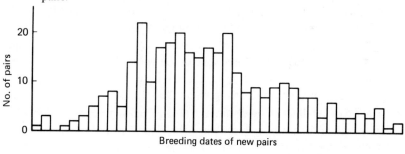

Breeding dates of new pairs

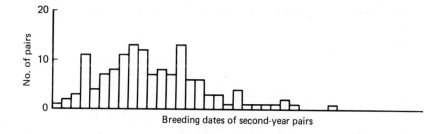

Breeding dates of second-year pairs

breed at a mean date of 22 June. From the mean breeding dates of table 8.3, we can estimate the time required for pair formation by dark, intermediate and pale males. This is given by the mean breeding dates of these males measured from 22 June, since 22 June is the date at which they would have bred if they had not had to find a new mate. Thus we calculate the times required for pair formation and mating:

Phenotype of male	Mean number of days required for pair formation
Dark	6.33
Intermediate	7.84
Pale	11.44

Pales take almost twice as long to find a new mate as dark males.

8.4.3 Male experience and pairing

If male competition determines the success in finding a mate, we would expect that males who have bred in previous years would be at an advantage over males who were breeding for the first time. If experience increases the effectiveness of courtship, experienced males should find mates more quickly than new males. But when a male is forced to find a new mate as the result of divorce or the death of his previous mate, his average breeding date is not significantly later than that of a male breeding for the first time. Table 8.4 shows the mean breeding dates of new and experienced males in new pairs and the analysis of variance of the data. Only the data for the period 1973–79 were used in the table. In these data there is no doubt which are the new birds. With very few exceptions all new breeding adults were colour-ringed in their first year in the colony. New, unringed birds were thus known to be breeding for the first time since there is no apparent movement of breeding birds from one colony to another.

The breeding dates of new pairs are always highly variable, unlike older pairs which show much reduced variance. The distributions in figure 8.1 show how much more variable the breeding dates of new pairs really are. This high level of intrinsic variability does give rise to large standard errors in the means. The data are barely significant ($P=0.047$) when analysed in terms of phenotypic differences alone.

When the orthogonal analysis of table 8.4 is used to test the significance of any interaction between experience and phenotype, no effect, not even of the differences between phenotypes, is found to exceed the level of significance $P=0.05$. Yet the means of new and experienced males do seem to be different: experienced intermediates breed slightly earlier than darks;

Table 8.4. *Breeding dates of new and experienced males in new pairs*

(i) Mean breeding dates of the males

Phenotype

Status	Dark	Intermediate	Pale	All phenotypes
New males	29.71	32.52	32.88	32.11
Experienced males	28.05	27.59	33.21	28.50

(ii) Analysis of variance in breeding dates

Source of variation	Sum of squares	Dfs	Mean square	Value of F	Value of P
Between new and experienced males	135.1512	1	135.15	2.495	0.117
Between phenotypes	302.7019	2	151.35	2.794	0.0639
Interaction	196.7662	2	98.383	1.817	0.166
Residual	9261.6762	171	54.162	—	—

new intermediates breed almost as late as pales. This difference could easily be produced by sampling variation, since the residual variance in new pairs is always so large. As we shall show in chapter 10, section 10.2, this high variance is an inevitable consequence of the process of mate selection. To get significant results, field work must be carried out over long periods of time. In the two periods of study on Fair Isle, breeding dates of 328 males in new pairs were collected over a total of 22 years. This very lengthy effort was indeed necessary to obtain the high overall levels of significance of the differences in breeding dates as shown in table 8.2.

8.4.4 Breeding dates of males and females

So far we have analysed the breeding dates of the males in relation to their phenotypes, breeding experience and whether they are breeding in new or older pairs. We have assumed that sexual selection operates, as it usually does, only between males. If the females were to show similar differences in breeding dates to those of the males, we should have to conclude that females were also subject to sexual selection. The mean breeding dates of males and females are compared in table 8.5 (i) and the analysis of variance is given in table 8.5 (ii). There is no doubt about the inference that may be drawn from the analysis of the data in table 8.5. When finding a new mate, the sexes clearly differ in their mean breeding dates. The females show less variation in their breeding dates than the males. They also vary in a different way to the males. The mean values

Table 8.5. *Breeding dates of males and females in new pairs*

(i) Mean breeding dates

		Phenotype		
Sex	Dark	Intermediate	Pale	All phenotypes
Male	28.33	29.74	33.44	30.07
Female	31.12	29.28	31.52	30.04

(ii) Analysis of variance in breeding dates

Source of variation	Sum of squares	Dfs	Mean square	Value of F	Value of P
Between sexes	2.163	1	2.163	0.03618	0.849
Between phenotypes	854.055	2	427.03	7.143	8.53×10^{-4}
Interaction	383.335	2	191.67	3.206	0.0411
Residual	39 575.954	662	59.782		

In this table 334 breeding dates of individual males and females have been used in the calculations. An additional six values have been added to the 328 values of table 8.1 on which the analyses of male differences have been based. These extra values have come from a number of unsexed new pairs of I × DI matings (I,-intermediate; DI, dark-intermediate).

Since both intermediates and dark-intermediates are heterozygous, they have all been classified as intermediate. An I × DI mating is therefore a mating between two heterozygotes. Such a mating supplies a breeding date for both a male and a female heterozygous intermediate without their having been sexed. In table 8.1 all matings of unsexed, like phenotypes were used to give breeding dates of males and females. But matings of the type I × DI were not originally classified as being of like phenotypes and so were not used to give breeding dates. All calculations on the males, including the extensive computer simulations to fit the models, were done without the six extra values for intermediates that could have been obtained from the unsexed I × DI matings. I decided not to re-calculate the fitting of the models after incorporating the six extra values, since the extra data make virtually no difference to the statistical significance of the variation between males.

suggest that intermediate females may be earlier breeders than the others. Consider the data of females alone: the differences between their mean breeding dates are not quite significant ($P=0.065$). But if, after inspection of the data, female intermediates (breeding at a mean 29.28 days after 1 June) are compared with darks and pales included together in one class (breeding at a mean 31.33 days after 1 June), then a significant result is found ($P=0.0202$). Since this was obtained only after inspection of the data and not by *a priori* hypothesis, the test is biassed and the difference cannot be judged significant.

The interaction in table 8.5 is only just significant at $P=0.0411$. The large residual variance, as we have already noted, makes the detection of interactions by statistical tests difficult unless the effects or the numbers of observations are very large. But although on the basis of this test alone, the difference between males and females may be considered somewhat doubtful, independent observations of behaviour prove that males and females are selected differently. Males usually keep the same territory after they have changed mates: females adopt the territories of the males they choose to mate with. A territory is primarily defended by the male, though with help from the female after mating. Courtship is also largely a male prerogative. On this evidence, females should not be subject to sexual selection: they should not differ in their breeding dates. The behavioural and statistical evidence justifies the inference that sexual selection acts on males but not on females.

8.5 Sexual selective values of Arctic Skua phenotypes

In the previous section we proved that the dark, intermediate, and pale males breed at different mean dates in a breeding season when they have to find new mates: pale males take five days longer to find mates than intermediate males; intermediates take slightly longer than darks. We must now ask: does this difference in the time to find a mate produce sexual selection? According to Darwin's theory, it will do so if later breeding reduces fledging success. We have already seen (section 7.4, chapter 7) that the reproductive success of all pairs, new and old, does indeed decline steeply towards the end of the breeding season. I obtained a general function relating relative fledging success to the successive weeks of the breeding season (O'Donald, 1972a). Analysis suggests that the fitness function holds regardless of the age of a pair in the colony: the same decline in breeding success is shown by all pairs – new pairs as well as old ones who have bred once or several times together in previous years. The individual values can be classified by the number of years a pair has bred in the colony and numbers of chicks they fledged.

For statistical analysis, the data were grouped into the weekly intervals shown in table 8.6. Old pairs usually breed early; very few breed towards the end of the season. Where there are only a few older pairs, these have to be lumped together with those of less experience as explained in the footnote to the table. If each pair, regardless of experience, fledges the same expected number of chicks in a given week, this expected number can be estimated by the average of the number of chicks fledged by all the pairs of that week. This value can be compared with the average number of chicks of pairs who have bred in the colony for 0, 1, 2, ... previous years. We can test the homogeneity of the values for pairs with differing previous

Table 8.6. *Fitness of pairs in relation to breeding date and years of breeding as a pair*

	Breeding dates in weekly intervals				
	11–17 June	18–24 June	25 June– 1 July	2–8 July	9–15 July
	1	2	3	4	5
Mean fitness in interval for all pairs	1.635	1.556	1.505	1.035	0.6875
Mean fitness of pairs *n* years together, *n* = 1	1.167	1.591	1.481	1.026	0.6552
2	1.667	1.446	1.676	0.818	1.000
3	1.600	1.421	1.400	1.375	—
4	1.750	1.581	1.125	—	—
5	1.556	1.700	—	—	—
6	1.667	1.727	—	—	—
7	1.833	1.875	—	—	—
8	1.800	1.750	—	—	—
9	1.600	—	—	—	—
Variance per pair	0.2858	0.3488	0.3601	0.4900	0.3398
Value of χ^2	6.799	9.430	6.528	2.949	0.951
Value of *P*	0.5585	0.2232	0.08856	0.2289	0.3295

Fitness is given by the average number of chicks fledged by a pair. Where values of fitness are missing, there are too few pairs to give good estimates. Data of such pairs were lumped with those with the largest number of years of breeding experience for which values are shown. For example, in week 3, pairs with more than four years' experience of breeding together were all included together with the pairs with just four years' experience.

experience by calculating χ^2 for the deviations from the overall weekly means. To calculate this χ^2, we must first calculate the variance of the average number of chicks. The standard χ^2 test, expressed by the formula

$$\chi^2 = \Sigma\{(\text{observed no.} - \text{expected no.})^2/(\text{expected no.})\}$$

cannot be used here, for the numbers of chicks are not simple counts of independent observations.

In terms of counting numbers of observations, the data take the form of the numbers of pairs fledging 0, 1 or 2 chicks. It would certainly be possible to set up contingency tables, classifying pairs according to their breeding experience together and whether they produced 0, 1 or 2 fledglings. The number of fledglings produced could not be used in such a test, however: this number is the value of an observation on a pair of birds, not the observation itself. Only numbers of observations can be used in the standard χ^2 test, which is based on the assumption that the numbers observed in any class have a variance equal to the numbers expected and are roughly Normal in distribution. However, we can calculate the mean and variance of the number of fledglings on the null hypothesis that in a particular week, a pair may produce 0, 1 or 2 fledglings, with probabilities p, q and r. We assume these probabilities are constant for all pairs, regardless of whether they have bred together before or how many years they have done so. Thus we have:

fitness, w 0   1   2

proportion of pairs p q r

$$\varepsilon(w) = q + 2r$$
$$V_w = q + 4r - (q + 2r)^2$$
$$= pq + qr + 4pr$$

where V_w is the variance in the number of fledglings a pair may produce. The probabilities p, q and r may be estimated by the total number of pairs that fledge 0, 1 or 2 chicks in a particular week. Then if w_i is the average number of fledglings produced by pairs who have i years of breeding experience together

$$\chi^2 = \Sigma(w_i - \varepsilon(w))^2 / V(w_i)$$

From the values of P for each week given in table 8.6 we then have

$$\chi^2 = \Sigma - 2 \log_e P$$
$$= 14.182$$

for 10 degrees of freedom since $- 2 \log_e P$ is distributed as χ^2 for two degrees of freedom. This is known as Fisher's 'Combination of probabilities' test and gives

$$P = 0.165$$

If the pairs do differ in loss of fitness through the breeding season, these data provide no evidence for any such difference.

From this analysis, we may conclude that the decline in fitness during the breeding season seems to affect all pairs equally, though of course pairs who have bred together before are usually the early ones. We can obtain a fitness function giving the general decline from the beginning to the end of the season, and use this to compare the average effect on fitness of the later breeding of pale males with the corresponding effect of the earlier breeding of dark males. Thus the sexual selective values can be estimated.

Table 8.7 *Relationship of fledging success to breeding date in the Arctic Skua*

(i) Chicks fledged by pairs in successive weeks of the breeding season

Dates in weekly intervals	No. of pairs with 0, 1 or 2 chicks			Total no. of pairs	Total no. of chicks	Mean fitness	Variance per pair
	0	1	2				
10–16 June	2	14	32	48	78	1.625	0.31771
17–23 June	9	73	136	218	345	1.583	0.32575
24–30 June	7	43	67	117	177	1.513	0.36949
1–7 July	13	32	21	66	74	1.121	0.50046
8–15 July	9	21	1	31	23	0.742	0.25598
15–21 July	4	3	1	8	5	0.625	0.48438

(ii) Goodness of fit of quadratic fitness function

Weeks of breeding season, x	Relative fitness, w $$w = 1 - \frac{(0.47354 - x)^2}{24.254}$$	Expected no. of chicks	Contribution to χ^2
0	0.99075	76.958	0.0712
1	0.98857	348.747	0.1977
2	0.90393	171.145	0.7928
3	0.73683	78.696	0.6678
4	0.48726	24.444	0.2627
5	0.15524	2.010	2.3076
	Total	702.000	4.300

Table 8.7 shows the fitness function that has been used for the calculation of the overall fitnesses of the dark, intermediate and pale males when they are mating with a new female. The details of the calculation of the function

$$w = 1 - \frac{(0.47354 - x)^2}{24.254}$$

are given by O'Donald (1972a). As table 8.7 shows, it is an excellent fit to the data of the numbers of chicks produced by pairs breeding in successive weeks of the season ($\chi^2 = 4.3$; $P = 0.37$).

Using the relative fitnesses of table 8.7 (ii) and the distributions of breeding dates of table 8.1, we find that in new pairs, intermediate males will produce 0.964 chicks for every one chick produced by a dark male. Pale males will produce 0.836 chicks for every dark male's chick. The relative selective disadvantage suffered by intermediate and pale males can be measured by selective coefficients S_I for intermediates and S_p for pales:

$$S_I = 0.036, \quad S_p = 0.164$$

We could equally well have calculated these selective coefficients using the empirical values of fitness given in table 8.7 (i). The values would then be

$$S_I = 0.039, \quad S_p = 0.162$$

The closeness of these two sets of values simply reflects the excellent fit of the quadratic function to the empirical values.

These selective coefficients apply only to males in new pairs. But new pairs represent only 38.6 per cent of all pairs. Therefore the effect of sexual selection on the population must be less than it would be if all matings were new every year. We must now estimate the selective coefficients for the population as a whole. Suppose the males in new pairs have reproductive rates as follows:

b_D for darks
b_I for intermediates
b_p for pales;

where these rates are determined by the average dates at which the males breed, and hence by sexual selection. In older pairs, let males all reproduce at rate B. The average reproductive rates of pale and dark males will then be given by

$$R_p = \{b_p n + B(N-n)\}/N$$
$$R_D = \{b_D n + B(N-n)\}/N$$

where n is the number of new pairs in a total of N pairs. The selective coefficient of pales, resulting from the difference of b_p and b_D, is therefore given by

$$s_p = 1 - R_p/R_D$$

$$= \frac{n}{N}\left(1 - \frac{b_p}{b_D}\right) \Big/ \left\{\frac{n}{N} + \frac{B}{b_D}\left(1 - \frac{n}{N}\right)\right\}$$

$$= \frac{n}{N}S_P \Big/ \left\{\frac{n}{N} + \frac{B}{b_D}\left(1 - \frac{n}{N}\right)\right\}$$

Similarly

$$s_I = \frac{n}{N}S_I \Big/ \left\{\frac{n}{N} + \frac{B}{b_D}\left(1 - \frac{n}{N}\right)\right\}$$

The data of table 8.7 consist of 169 new pairs which produced 212 fledged chicks and 319 older pairs which produced 490 fledged chicks. Thus we estimate

$$B = 490/319$$
$$= 1.536$$

Since $n/N=0.386$ and $b_D=1.334$, we obtain the overall coefficients of sexual selection

$$s_I = 0.0128, \quad s_p = 0.0579$$

These values are slightly different from those given in a previous paper (O'Donald, 1980c). In that paper I applied the approximate formula

$$s_I \approx \frac{n}{N}S_I \qquad s_p \approx \frac{n}{N}S_p$$

and obtained slightly larger estimates than the estimates given here. The approximate formula is quite satisfactory in view of the likely magnitude of the standard errors of the estimates. But, although the selective coefficients are certainly very significant – they are derived from the highly significant differences in the distributions of breeding dates – their standard errors could only be calculated very roughly using a series of approximations. I decided this calculation was not worth attempting.

It is well to remember that the calculation of these sexual selective coefficients is based solely on the distribution of breeding dates of the males in new pairs. This assumes that the differences in the males' breeding dates are caused by sexual selection. This assumption is based on evidence that only males in new pairs show the differences: neither the females nor males in older pairs differ significantly in their breeding dates; in this respect males in new pairs are significantly different from both the females and the males in older pairs. In new pairs, intermediate and pale males breed somewhat later on average than dark males. The selective coefficients of the intermediate and pale males relative to the darks are then calculated by assuming that later breeding reduces fledging success, as originally postulated by Darwin. And this assumption is based on evidence that all pairs, both newly formed and older, suffer a similar reduction in fitness late in the breeding season.

In this chapter I have not postulated any mechanism for the sexual selection. Dark males may compete more fiercely for the females, or the females may prefer to mate with them. Either way, a dark male would have a better chance of mating. At each stage in the breeding season, a dark male would be more likely to find a mate. On average he will breed somewhat earlier in the season than the others, and thus gain his reproductive advantage by fledging more chicks.

Different behavioural mechanisms of sexual selection have different effects both on the distributions of breeding dates and frequencies of matings and on the long-term evolutionary outcome of selection. In the following chapter, I analyse the genetic consequences of different possible mechanisms of sexual selection. In chapter 10, behavioural models of mate selection are fitted to the Arctic Skua data. Thus the parameters of the models – the mating preferences, for example – can be estimated and the goodness of fit of the models tested.

9

Genetic models of sexual selection in birds

9.1 Female preferences and male competition in monogamous birds

Let us suppose that some males possess a genetic characteristic that can increase female arousal. Some females, responding more rapidly to these males, thus mate with them preferentially. Other females, however, show no preferences: they just choose any male at random. To construct a theoretical model of the sequence of matings, I assume that groups of females come into breeding condition during successive intervals of the breeding season. They choose their mates from the available pool of unmated males. This pool decreases, of course, as each successive group of females choose their mates. Later females may not be able to mate with the males they prefer because none may be left in the pool.

In monogamous species, the sequence of matings is clearly important in determining what choices can be made. This applies to the matings taking place within a particular group of individuals. Females with particular preferences might choose their mates before, at the same time as, or after the females who mate at random. If preferential matings come first, more females will be able to mate with the males they prefer. But if the random matings come first, some preferred males will already have been chosen when the preferential matings take place: females will then have less opportunity to express their preferences.

The behavioural mechanism of mate selection will presumably determine the sequence of preferential and random matings. Females may differ in their responses to the males: males may differ in the intensity and persistence of their courtship. Preferential mating may follow in each case. Females presumably require the stimulation of male courtship to elicit their mating behaviour: after receiving a certain amount of stimulation, a threshold is reached at which a female can respond. No doubt this threshold varies between different individuals. Some females will inevitably be slower to respond than others. Certainly, in *Drosophila*, strains differ genetically in their response to stimulation (Ehrman & Spiess, 1969). In some strains, females are ready to mate after only a brief encounter with a courting male; in other strains, the females are much more reluctant to mate and need the stimulation of much more courtship before they can respond. Such variation in responsiveness is, no doubt, universal in animals that use courtship for attracting the opposite sex. Variation in the intensity and duration of courtship will then give rise to some degree of preferential mating. Males who more actively or persistently court the females will be the more likely to get a response, particularly from the less responsive females. Encounters with courting males will raise a female's level of stimulation. She finally mates with the male who stimulates her above her threshold. A more active and persistent male will be more likely to succeed in mating with her.

Females with low thresholds requiring little courtship before responding will obviously mate before the females with high thresholds. If any male can stimulate a responsive female to mate, she will tend to mate at random. Less responsive females, who must receive more courtship before being sufficiently stimulated to respond, will tend to show preferences in favour of more active and persistent males. They will also be the last females to choose their mates. The more responsive females will thus tend to mate randomly and sooner than the less responsive females: random matings will generally precede preferential matings. Of course, a less responsive female may by chance encounter only the less active males. After a sufficient number of encounters, she will eventually mate with one of them. Whether she exercises a preference in favour of the more active males will depend on the chance she encounters one of them before her threshold level of stimulation is reached – hence on their frequency in the population. If genetic factors determine levels of male courtship, the probability of encounter increases as the females select the more active males which thus increase in frequency in later generations. Since in general, the number of encounters depends on the different thresholds of the females, this, together with male frequency, determines the probabilities that females express a preference. Detailed encounter models of sexual selection have been formulated to describe the process of preferential mating in polygynous species (O'Donald, 1978, 1979, 1980a; Karlin & Raper, 1979; Raper, Karlin & O'Donald, 1979). They give rise to frequency-dependent expression of preference. This in turn determines the degree of frequency-dependence in the number of matings achieved by the males. Preferential matings always produce some frequency-dependence in mating advantage such that rare males usually mate more often than common males. This is the 'rare-male effect' that gives rise to genetic polymorphism when several phenotypes may be the objects of preference or when sexual selection is balanced by natural selection (see section 8.3).

The process I have described produces preferential mating in favour of sexually more active or persistent males. No conscious choice is being ascribed to the females. They are only responding sooner or later and at different thresholds to the males' courtship. It may be argued that this is female choice but not female preference: the females have no prior preference but choose the males by the intensity and persistence of their courtship. This is merely a semantic point. I use the term 'mating preference' descriptively whenever females are shown to differ in the relative frequencies at which they choose between different male phenotypes or genotypes. It is easy to imagine how genetic differences in courtship might arise. Aggressive behaviour, territorial behaviour and courtship behaviour are closely linked to gonadotrophin and androgen levels in males. A male that is more aggressive at the beginning of the breeding season driving off other males and maintaining a larger territory, may also court the females more vigorously later. Male competition and female choice both contribute to his success.

There is evidence that melanic pigeons gain a mating advantage by increased courtship activity. Murton *et al.* (1973) found that the melanic, blue-checker and dark-blue-checker males have larger testes and higher sperm counts than the non-melanic, wild-type males. The melanics' gonads recrudesce earlier in the spring and are less likely to regress in the autumn: they have an altered photoperiod and longer breeding season. More recent work has shown (J.M. French, unpublished Ph.D. Thesis) that melanics show a more rapid rise in gonadotrophin on being transferred to a longer photoperiod and produce higher levels of androgen. These differences are associated with greater courtship activity in the head bobbing display to females. This appears to produce a very great mating preference in favour of the melanic phenotypes. Murton *et al.* had previously observed matings between a number of different phenotypes in a population of pigeons in Salford containing a high frequency of melanics. The matings actually observed are largely disassortative: wild-types strongly preferred mating with blue-checkers; blue-checkers preferred dark-blue-checkers or wild-types. By fitting a detailed model of disassortative mating, Davis & O'Donald (1976*b*) showed that 51 per cent of the females preferred blue-checker and 38 per cent preferred dark-blue-checker. Almost 90 per cent of the females had preferences they would express provided they themselves did not have the phenotypes they preferred. It is plausible to explain these remarkable findings by the earlier breeding and more active courtship of the melanic males coupled with inhibition of female response when the courting male has the same phenotype as the female. The estimates of the female preferences certainly fit this interpretation very well.

So far I have explained the expression of female preference in terms of response to different levels of male courtship. A female responds at a particular threshold in the total amount of stimulation she has received. She responds at the same threshold to all males. She always needs the same total amount of stimulation; but some males are more active and more persistent than others – for example, the melanic pigeons. They are more likely to give the females the extra stimuli needed for response. Hence the females mate preferentially with the 'sexier' males. The less responsive the female, the more stimuli she needs and the greater the probability she mates with a sexier male. The expression of female preference can also be explained by variation in the threshold of female response (O'Donald 1979, 1980*a*). Some females may respond at one threshold when courted by one phenotype of male and at another threshold when courted by another phenotype of male. The males themselves would not necessarily differ in the levels of their courtship activity, though they will show polymorphism in the phenotypes which determine the thresholds of female response.

The red throat of the male stickleback provides a clear example of this mechanism of expressing preference. The development of the male's red throat in the breeding season is a genetically determined character. Some populations are polymorphic for red and non-red males. Semler (1971) carried out experiments offering females a choice of laying their eggs in nests of red or non-red males. The eggs are fertilized by the male who entices the female into his nest. Aquaria were divided by glass partitions into two halves. A red and non-red male built their nests and established their territories each in their own half. The partition was removed and a gravid female introduced. She chose one or other male and laid her eggs in his nest. The red males were strongly preferred. At the same time, it was demonstrated that the preference was not exercised in favour of red males because they were the more active in courting the females. Non-red males with a red throat painted on with lipstick or nail varnish were just as effective in securing a mating as naturally red-throated males. A large proportion of the females were simply 'turned on' by the males' red throats. This mechanism is what one would normally understand by female preference: some females are attracted by a particular male characteristic and choose males who have it. There is difference both in the males and in the females' responses to them. Some such mechanism must be invoked to explain the evolution of female preference. According to Fisher's theory (see chapter 8, section 8.2), females evolve a preference for a male characteristic which is pleiotropically associated with, or genetically linked to, characters for increased fitness. The preferred character acts as a phenotypic marker by which the fitter males can be discriminated. A preference can be imagined to evolve if, genetically, some females have a lower threshold of response towards the males with the marked phenotype. These females can then gain selective advantage by producing fitter sons.

Female preference could equally well evolve by the selection of females with a genetically higher threshold towards males who lack the marked phenotype. Instead of being more readily turned on by the male stickleback's red throat, a female could equally well express a preference for red males if non-red males turned her off. These alternatives seem to be effectively identical. But they would have different consequences on the sequence in which the matings took place. If some females exercise a preference because they have a lower threshold of response to the preferred males, preferential matings will precede random matings. If, on the other hand, females with preferences have a higher threshold to all males except those with the preferred phenotypes, preferential matings will occur simultaneously with random matings, since they take place at the same average threshold level of stimulation both in females with preferences and in females without preferences.

As we have seen, it is usually advantageous to mate sooner rather than later in the breeding season. The evolution of higher thresholds will delay the average date of mating. This will produce selective disadvantage of females with raised thresholds. Preferences are much more likely to evolve by selection of females with lower thresholds towards preferred males. In a sequence of matings, we should thus expect that preferential matings would precede random matings. But if females respond at the same thresholds to all males, while males differ in the intensity and persistence of their courtship, then we should expect a sequence in which random matings would tend to precede preferential matings; for the less responsive females, mating later, would be the more likely to mate with the more active, and hence more stimulating, of the males.

In the Arctic Skua, I have no evidence based on direct observation of mating behaviour to indicate that melanic and non-melanic birds differ in their courtship. Assortative mating of melanic with melanic phenotypes suggests mate selection takes place by female choice. It might take place by a similar process to that in pigeons, except that pigeons mate disassortatively, not like with like. Melanic males might be the more active in courting the females, while females' responses are inhibited if their phenotype is unlike the males. Unfortunately it is very difficult to observe the details of the Arctic Skua's courtship. It is mainly aerial, with aerobatics and high-speed chases, and very difficult to follow and analyse. After pairing, the birds seem to do very little, sitting or standing in their territories for long periods, occasionally taking off to intercept intruders and chasing them off after a furious dog-fight. A pair copulates several times, but copulation is seldom preceded by much obvious courtship (except for 'begging' as described in section 7.1.3. chapter 7). The stimulus to copulate often seems to be the copulation of a pair in an adjacent territory.

The mechanism of sexual selection in the Arctic Skua must be inferred by less direct methods than observation of the process of mate selection. Models of mate selection can be used to predict the frequencies of matings and their sequences in relation to breeding dates. The models can be tested for goodness of fit to the data of the Arctic Skua. Some of these data have already been used in evidence of sexual selection – the data of breeding dates, for example. In chapter 10, these data will be used for fitting the models and estimating parameters. Models of assortative mating can be fitted to data on frequencies of matings between phenotypes. The rate and final outcome of evolution is different in each model. Since the genetic structure of Arctic Skua populations is a product of past evolution, it can be used to infer the nature of the evolutionary processes that gave rise to it. To deduce the evolutionary consequences of different mechanisms of sexual selection is thus my purpose in describing and analysing models of sexual selection in the following sections of this chapter.

9.2 Genetic models of preferential mating

There are two ways of setting up genetic models of preferential mating in the Arctic Skua. One is to use a computer to simulate the matings that take place as female Arctic Skuas come into breeding condition and choose their mates according to their preferences. The actual number of Arctic Skuas breeding in a particular period or interval of the season can be considered to represent a group of females who are choosing their mates. This number of females can be used in the computer simulation to give the numbers of dark, intermediate and pale males the females mated with. In any particular interval of the breeding season, a pair will produce a certain average number of fledglings (table 8.7, section 8.5, chapter 8). The fledging success of dark, intermediate and pale males can thus be calculated for each each interval and hence their fitnesses averaged over the whole season. For example, the data of table 8.7 were used in some of the simulations, the breeding season having been divided into the weekly intervals shown in the table. In other simulations, the intervals were successive days, the numbers of females mating each day and the fledging success having been obtained by fitting hypothetical distributions (O'Donald, 1973). The computations are obviously complicated and cannot possibly be stated in terms of explicit equations that might be solved algebraically.

Genetic models can also be set up by simplifying the sequence of events during the breeding season. But a simpler model may still be biologically realistic. As a result of the complicated sequence of events taking place during the breeding season, the preferential matings take place on average sooner and produce more offspring than the random matings. The models can be simplified by assuming that all preferential matings take place first at one level of fertility, followed by all the random matings at another, lower level of fertility. O'Donald (1980*b*) analysed models with three different modes of expression of the females' preferences: (i) females preferred phenotypes A or a, of which A is dominant to a; (ii) females preferred genotypes AA, Aa or aa; (iii) females preferred either a dominant or a recessive phenotype which reduced the chance of survival, natural selection thus opposing sexual selection.

9.2.1 Preferences for two phenotypes with dominance

Whenever monogamous species mate preferentially, some females may not always be able to find the males of their choice: the males they prefer may all have been mated; more females may have preferences than there are males they prefer. Some females will then be disappointed. Eventually we may assume they will mate with any male who is still unmated. In most of the models, I assume preferential matings precede random matings. For the reasons given in the previous section, I consider that females mate preferentially because they respond at a lower threshold to the males they prefer. The random matings take place later after a higher threshold has been reached.

In the model, I assume that the two phenotypes, A (dominant) and a (recessive), occur in a population at relative frequencies 1-w and w. These phenotypes can be considered to represent the melanic (dark and intermediate) and non-melanic (pale) phenotypes of the Arctic Skua. The melanics are at least semi-dominant: they are similar in appearance and overlap to some extent in the expression of their dark plumage. A consists of the genotypes AA (dark) and Aa (intermediate) at frequencies u and v; a

consists of the genotype *aa*. The melanic and non-melanic males are preferred by proportions α and β, respectively, of the females.

Three cases must be considered: Case (i), in which $\alpha \le 1 - w$, $\beta \le w$, so that all females with preferences can mate with the males they prefer; Case (ii), in which $\alpha \le 1 - w$, $\beta > w$, so that too many females prefer *a* males; and Case (iii), in which $\alpha > 1 - w$, $\beta \le w$, so that too many females prefer *A* males. A female who prefers *A* males may choose either *AA* or *Aa* indiscriminately, with probabilities $u/(1 - w)$ and $v/(1 - w)$.

Case (i). Assume that the conditions for Case (i) hold. Males then take part in matings with their genotypes at frequencies

	in preferential matings	in random matings
AA	$\alpha u/(1 - w)$	$u - \alpha u/(1 - w)$
Aa	$\alpha v/(1 - w)$	$v - \alpha v/(1 - w)$
aa	β	$w - \beta$

Matings between different genotypes occur with frequencies and fertilities shown in table 9.1. It has been assumed that the random matings have a fertility $f (1 \ge f \ge 0)$ compared with the fertility 1.0 of the preferential matings. The genotypic frequencies in the next generation can then be derived as follows from the fertilities of the matings and the offspring produced:

$$Tu' = fp^2 + \alpha(1 - f)p^2/(1 - w)$$
$$Tv' = 2fpq + \alpha(1 - f)(2pq - pw)/(1 - w) + \beta(1 - f)p$$
$$Tw' = fq^2 + \alpha(1 - f)(q^2 - qw)/(1 - w) + \beta(1 - f)q$$

where $T = f + (1 - f)(\alpha + \beta)$ as shown in the table. It is obvious, as we should expect, that when $\alpha = \beta = 0$, simple random mating takes place; the genotypic frequencies then follow the Hardy-Weinberg law:

$$u' = p^2$$
$$v' = 2pq$$
$$w' = q^2$$

The gene frequency in the next generation is given by
$$p' = u' + \tfrac{1}{2}v'$$
$$= p$$

and has therefore remained unchanged. No selection occurs: the gene and genotype frequencies stay the same in all subsequent generations.

Similarly, when both the preferential and the random matings have the same fertilities, with $f = 1$, the matings are effectively random and the

Table 9.1. *Frequencies of preferential and random matings and the fertilities of matings as determined by preferences for both dominant and recessive phenotypes*

(i) Frequencies of matings

Mating	Preferential component of frequency	Random component of frequency
$AA \times AA$	$\alpha u^2/(1-w)$	$u^2 - \alpha u^2/(1-w)$
$AA \times Aa$	$2\alpha uv/(1-w)$	$2uv - 2\alpha uv/(1-w)$
$AA \times aa$	$\beta u + \alpha uw/(1-w)$	$2uw - \beta u - \alpha uw(1-w)$
$Aa \times Aa$	$\alpha v^2/(1-w)$	$v^2 - \alpha v^2/(1-w)$
$Aa \times aa$	$\beta v + \alpha vw/(1-w)$	$2vw - \beta v - \alpha vw/(1-w)$
$aa \times aa$	βw	$w^2 - \beta w$

(ii) Fertilities of matings and offspring produced

Mating	Fertility	Offspring AA	Aa	aa
$AA \times AA$	$[fu^2 + \alpha(1-f)u^2/(1-w)]/T$	1	—	—
$AA \times Aa$	$[2fuv + 2\alpha(1-f)uv/(1-w)]/T$	$\frac{1}{2}$	$\frac{1}{2}$	—
$AA \times aa$	$[2fuw + \alpha(1-f)uw/(1-w) + \beta(1-f)u]/T$	—	1	—
$Aa \times Aa$	$[fv^2 + \alpha(1-f)v^2/(1-w)]/T$	$\frac{1}{4}$	$\frac{1}{2}$	$\frac{1}{4}$
$Aa \times aa$	$[2fvw + \alpha(1-f)vw + \beta(1-f)v]/T$	—	$\frac{1}{2}$	$\frac{1}{2}$
$aa \times aa$	$[fw^2 + \beta(1-f)w]/T$	—	—	1

$$T = f + (1-f)(\alpha+\beta) = 1 - (1-f)(1-\alpha-\beta)$$

Hardy-Weinberg law again holds as above. Some reduction in the average fertility of the random matings is of course necessary for selection to occur.

In general, therefore, the gene frequency of A in the next generation is given by

$$Tp' = fp + \alpha p(1-f)(1-\tfrac{1}{2}w)/(1-w) + \tfrac{1}{2}\beta p(1-f)$$

If the gene frequency has changed between generations by the amount

$$\Delta p = p' - p$$

then

$$T\Delta p = \tfrac{1}{2}p(1-f)[w(\alpha+\beta) - \beta]/(1-w)$$

There will be a point of equilibrium when $\Delta p = 0$. At this point, $p_* = 0$, or

$$w_* = \beta/(\alpha+\beta)$$

The point $p_* = 0$ merely represents the case when no mutation A has occurred in the population. But the point $w_* = \beta/(\alpha+\beta)$ represents a polymorphic equilibrium at which the phenotypes A and a are both maintained in the population at frequencies $1 - w_*$ and w_*. It is easy to show that at this equilibrium, and only at this equilibrium, the Hardy-Weinberg ratios hold:

$$u_* = p_*^2$$
$$v_* = 2p_*q_*$$
$$w_* = q_*^2$$

Therefore the allele a has its equilibrium at the gene frequency

$$q_* = [\beta/(\alpha+\beta)]^{\frac{1}{2}}$$

while

$$p_* = 1-q_* = 1-[\beta/(\alpha+\beta)]^{\frac{1}{2}}$$

It is not sufficient simply to demonstrate that equilibrium points may exist. They may be either stable or unstable points. Populations move towards stable points and away from unstable points. To determine the outcome of an evolutionary process it is necessary to determine whether equilibria are stable or not. For the present case, it is sufficient to note that the difference equation representing the change in gene frequency may be put in the form

$$T\Delta p = \tfrac{1}{2}p(1-f)\,(w-w_*)\,(\alpha+\beta)/(1-w)$$

Suppose $w > w_*$. Since all other terms are positive, this entails $\Delta p > 0$, and hence an increase in the frequency of A. Therefore a declines in frequency and w approaches w_*. Similarly $w < w_*$ entails $\Delta p < 0$, and A declines while a increases. Again w approaches w_*. The values of w always therefore converge on the equilibrium value w_*. The equilibrium point w_* is said to be 'globally stable', since the population will always end up there, regardless of where it started from. As the equilibrium is approached, the difference of the gene frequency p from the equilibrium frequency p_* is reduced by a constant factor every generation. Suppose p_0 is a small deviation from p_*. After n generations, the gene frequency p_n will be much closer to p_*. O'Donald (1980b) showed that

$$p_n-p_* = \lambda^n(p_0-p_*)$$

where

$$\lambda = 1 - \frac{p_*q_*(1-f)\,(\alpha+\beta)^2}{\alpha[f+(1-f)\,(\alpha+\beta)]}$$

Since $\lambda < 1$ for $\alpha, \beta > 1$, this shows that $p_0, p_n \rightarrow p_*$. The rate of convergence is said to be 'geometric', since the deviation $p-p_*$ diminishes by the factor λ every generation. The condition $\lambda < 1$, which is necessary for convergence, holds only if $f < 1$, which is the assumption of Darwin's model of sexual selection in monogamous birds. If $f = 1$, then of course, nothing happens:

pure random mating prevails since all individuals eventually mate and no one suffers any loss of fertility. If $f > 1$, the later matings would be the more fertile. This would obviously lead to $\lambda > 1$. Now the deviations from equilibrium will get larger, rather than smaller, every generation. This shows that the equilibrium point

$$w_* = \beta/(\alpha+\beta)$$
$$p_* = 1-[\beta/(\alpha+\beta)]^{\frac{1}{2}}$$

would be unstable. Either A or a would ultimately be eliminated depending on whether p was less or greater than p_*.

These results may be compared with the effects of sexual selection in a polygynous species. Then it can be shown that

$$w_* = \beta/(\alpha+\beta)$$
$$p_* = 1-[\beta/(\alpha+\beta)]^{\frac{1}{2}}$$

with geometric rate of convergence given by

$$\lambda = 1-p_*q_* (\alpha+\beta)^2/\alpha$$

The equilibrium is the same, but the rate of convergence is always much faster with polygyny. One interesting result is that the final equilibrium in a monogamous species is independent of f, provided that $f < 1$. Only at the point $f = 1$ does the initial gene frequency remain unchanged. Selection can be very slow if $f \approx 1$, because the factor $1-f$ determines the rate of approach.

Case (ii). According to the conditions for Case (ii), $\alpha \leq 1-w$, $\beta > w$, so that too few a males are available to mate with all the females that prefer them. The males thus mate with the frequencies

	in preferential matings	in random matings
AA	$\alpha u/(1-w)$	$u-\alpha u/(1-w)$
Aa	$\alpha v/(1-w)$	$v-\alpha v/(1-w)$
aa	w	0

giving the recurrence equations

$$Tu' = p^2-p^2(1-f)\,(1-\alpha-w)/(1-w)$$
$$Tv' = 2pq-(1-f)\,(2pq-pw)\,(1-\alpha-w)/(1-w)$$
$$Tw' = q^2-(1-f)\,(q^2-qw)\,(1-\alpha-w)/(1-w)$$

so that

$$Tp' = p-p(1-f)\,(1-\tfrac{1}{2}w)\,(1-\alpha-w)/(1-w)$$

and

$$T\Delta p = -\tfrac{1}{2}pw(1-f)\,(1-\alpha-w)/(1-w)$$

where

$$T = f+(1-f)(\alpha+w) = 1-(1-f)(1-\alpha-w)$$

Thus equilibrium is apparently reached when

$$w_* = 1-\alpha$$

But, by the conditions for this case, $\beta > w$, and hence at equilibrium $\beta > 1-\alpha$ so that we should have

$$\alpha+\beta > 1$$

which contradicts the essential premise of the model that $\alpha+\beta \le 1$.

The frequency $w_* = 1-\alpha$ cannot represent the equilibrium point. Before this point can be reached, the increase in w must give rise to the condition $\beta \le w$. The conditions for Case (ii) thus become those of Case (i) in the course of evolution. Therefore the equilibrium point of Case (i) is ultimately reached.

At the start of selection, either A or a must be at low frequency having entered the population by migration or mutation. If a is low in frequency, the conditions for Case (ii) will hold: a increases until the conditions change to those of Case (i), and the final outcome is the equilibrium of Case (i). If A is low in frequency, having just entered the population, then the conditions of Case (iii) will hold initially.

Case (iii). The conditions $\alpha > 1-w$, $\beta \le w$ show that too few A males are available to mate with all the females that prefer them. The males mate with frequencies

	in preferential matings	in random matings
AA	u	0
Aa	v	0
aa	β	$w-\beta$

giving the recurrence equations

$$Tu' = p^2$$
$$Tv' = 2pq-p(1-f)(w-\beta)$$
$$Tw' = q^2-q(1-f)(w-\beta)$$

and

$$Tp' = p-\tfrac{1}{2}p(1-f)(w-\beta)$$

so that

$$T\Delta p = \tfrac{1}{2}p(1-f)(w-\beta)$$

Table 9.2. *Equilibrium frequencies and rates of approach to equilibrium in models with dominance and either polygynous or monogamous mating*

Model	Equilibrium frequency	Geometric rate of approach to equilibrium
Equal fertility for all matings		
Monogamy	$w_* = q_0^2$	$\lambda = 1$
Polygyny	$w_* = \beta/(\alpha+\beta)$	$\lambda = 1 - p_*q_*(\alpha+\beta)^2/\alpha$
Fertility reduced in later matings		
Monogamy	$w_* = \beta/(\alpha+\beta)$	$\lambda = 1 - \dfrac{p_*q_*(1-f)\,(\alpha+\beta)^2}{\alpha[f+(1-f)\,(\alpha+\beta)]}$
Polygyny	$w_* = \beta/(\alpha+\beta)$	$\lambda = 1 - \dfrac{p_*q_*(\alpha+\beta)^2}{\alpha[f+(1-f)\,(\alpha+\beta)]}$

This equation implies that at equilibrium

$$w_* = \beta$$

which again entails the impossible condition

$$\alpha+\beta > 1$$

But just as with Case (ii), the conditions change to those of Case (i) before the impossible equilibrium has been reached. The final equilibrium is always that of Case (i). Table 9.2 summarizes the results of the analysis of the dominance model both for monogamous and polygynous species.

9.2.2. Separate preferences for each of the genotypes

Although dark and intermediate Arctic Skuas resemble each other much more closely than either resemble the pales, they are recognizably distinct: the intermediates have white bases to the belly feathers, giving a somewhat paler appearance when compared to darks. They also have the pale or lighter brown collar round the neck. It is possible that the genotypes for dark, intermediate and pale may be the objects of separate preferences. Suppose AA are dark, Aa intermediate and aa pale. When females may prefer one of these three genotypes, three parameters, α, β and γ, are required for the proportions of females who prefer AA, Aa or aa males. There are now seven possible sets of initial conditions:

Case (i), $\alpha \le u,\ \beta \le v,\ \gamma \le w$;
Case (ii), $\alpha \le u,\ \beta \le v,\ \gamma > w$;

Case (iii), $\alpha \leq u$, $\beta > v$, $\gamma \leq w$;
Case (iv), $\alpha > u$, $\beta \leq v$, $\gamma \leq w$;
Case (v), $\alpha \leq u$, $\beta > v$, $\gamma > w$;
Case (vi), $\alpha > u$, $\beta \leq v$, $\gamma > w$;
Case (vii), $\alpha > u$, $\beta > v$, $\gamma \leq w$.

Cases (ii) and (iv) and (v) and (vii) should give essentially similar results in view of their symmetries.

Case (i). All females with preferences can mate with the males they prefer. The males therefore mate with the frequencies

	in preferential matings	in random matings
AA	α	$u - \alpha$
Aa	β	$v - \beta$
aa	γ	$w - \gamma$

The matings then take place with fertilities as shown in table 9.3. We can see immediately that if $f = 1$, or $\alpha = \beta = \gamma = 0$, the fertilities simply equal the frequencies of random mating. The recurrence equations for the genotypes are quickly obtained and hence the recurrence equation for the gene frequency

$$Tp' = p - (1 - f)[p(1 - \tfrac{1}{2}\alpha - \tfrac{1}{2}\beta - \tfrac{1}{2}\gamma) - \tfrac{1}{2}\alpha - \tfrac{1}{2}\beta]$$

This is a simple linear equation in one variable. It gives the difference equation

$$T\Delta p = \tfrac{1}{2}(1 - f)\,[\alpha + \tfrac{1}{2}\beta - p(\alpha + \beta + \gamma)]$$

showing that

$$p_* = (\alpha + \tfrac{1}{2}\beta)/(\alpha + \beta + \gamma)$$

The Hardy-Weinberg Law holds at this equilibrium, exactly as when matings are polygynous.

If we put $\alpha + \tfrac{1}{2}\beta = \phi$ and $\alpha + \beta + \gamma = \theta$, then the recurrence equation can be written in the form

$$p' = p\left[\frac{f + \tfrac{1}{2}\theta(1 - f)}{f + \theta(1 - f)}\right] + \frac{\tfrac{1}{2}\phi(1 - f)}{f + \theta(1 - f)}$$

with the general solution for all values of p

$$p_n - p_* = (p_0 - p_*)\left[\frac{f + \tfrac{1}{2}\theta(1 - f)}{f + \theta(1 - f)}\right]^n$$

Table 9.3. *Fertilities of matings with separate preferences for each genotype*

Mating	Fertility
$AA \times AA$	$\alpha u + fu(u-\alpha) = u^2 - u(1-f)(u-\alpha)$
$AA \times Aa$	$\alpha v + \beta u + fv(u-\alpha) + fu(v-\beta) = 2uv - (1-f)(2uv - \alpha v - \beta u)$
$AA \times aa$	$\alpha w + \gamma u + fw(u-\alpha) + fu(w-\gamma) = 2uw - (1-f)(2uw - \alpha w - \gamma u)$
$Aa \times Aa$	$\beta v + fv(v-\beta) = v^2 - v(1-f)(v-\beta)$
$Aa \times aa$	$\beta w + \gamma v + fw(v-\beta) + fv(w-\gamma) = 2vw - (1-f)(2vw - \beta w - \gamma v)$
$aa \times aa$	$w + fw(w-\gamma) = w^2 - w(1-f)(w-\gamma)$

In this table the fertilities must be divided by the factor
$$T = 1 - (1-f)(1-\alpha-\beta-\gamma)$$
to bring the total fertility to 1.0.

Table 9.4. *Equilibrium frequencies and rates of approach to equilibrium in models with separate preferences for each genotype and either polygynous or monogamous mating*

Model	Equilibrium frequency	Geometric rate of approach to equilibrium
Equal fertilities for all matings		
Monogamy	$p_* = p_0$	$\lambda = 1$
Polygyny	$p_* = \phi/\theta$	$\lambda = 1 - \frac{1}{2}\theta$
Fertility reduced in later matings		
Monogamy	$p_* = \phi/\theta$	$\lambda = \dfrac{f + \frac{1}{2}\theta(1-f)}{f + \theta(1-f)}$
Polygyny	$p_* = \phi/\theta$	$\lambda = \dfrac{f + \frac{1}{2}\theta(1-2f)}{f + \theta(1-f)}$

In this table $\phi = \alpha + \frac{1}{2}\beta$ and $\theta = \alpha + \beta + \gamma$

In the polygynous model, the corresponding equation is
$$p_n - p_* = (p_0 - p_*)(1 - \tfrac{1}{2}\theta)^n$$
As in the model with dominance, the equilibrium is stable for all values $f < 1$. The results of selection in monogamous and polygynous species are compared in table 9.4

When gene frequency is increasing in a population from a low initial value, the conditions will first be those for cases (v) or (vii). At some point, they will change to the conditions for cases (ii), (iii) or (iv), and then to the conditions for case (i), giving the final equilibrium at the frequency

$$p_* = (\alpha + \tfrac{1}{2}\beta)/(\alpha + \beta + \gamma)$$

But unlike the dominance model, this does not always happen. Sometimes the final equilibrium can be attained while the conditions for cases (ii), (iii) or (iv) still hold and before the equilibrium of case (i) has been attained. Detailed analyses of all cases of this model have been published in an earlier paper (O'Donald, 1980b). In view of the estimated parameter values, only case (i) is applicable to the Arctic Skua, with equilibrium given above. In this case, as with all cases of dominance, the final equilibrium is just the same as in models of polygynous sexual selection. Only at the value $f = 1$ does no selection take place, the population remaining at Hardy-Weinberg proportions with the initial gene frequencies.

9.2.3 *Preference for a dominant phenotype with natural selection against it*

We have already found that natural selection acts to the disadvantage of the sexually favoured dark and intermediate phenotypes as a result of their later onset of maturity (chapter 6, section 6.3).

Characters that have evolved by sexual selection must often reduce the chances of survival, since an elaborate display to attract the females may also attract predators. The development of heavy plumage and other characters for display may also hinder escape. This natural selection can be measured by selective coefficients, s and t, where s measures the relative disadvantage of the dominant phenotype and t the relative disadvantage of the recessive. After natural selection against the recessive, the male phenotypes A and a occur at frequencies

$$[A] = (1-w)/(1-tw)$$
$$[a] = w(1-t)/(1-tw)$$

The single coefficient t is in fact sufficient to measure both advantage and disadvantage if negative values are allowed. Thus, if t is negative, a is advantageous compared to A. This is equivalent to using s to measure the disadvantage of A, since, after selection against A,

$$[A] = (1-s)(1-w)/[1-s(1-w)]$$
$$[a] = w/[1-s(1-w)]$$

and therefore

$$t = -s/(1-s)$$

The substitution of $-s/(1-s)$ for t converts selection against the recessive

Table 9.5. *Frequencies of preferential and random matings and the fertilities of matings as determined by preferences for both dominant and recessive phenotypes*

(i) Frequencies of matings

Mating	Preferential component of frequency	Random component of frequency
$AA \times AA$	$\alpha u^2/(1-w)$	$u^2/(1-tw)-\alpha u^2/(1-w)$
$AA \times Aa$	$2\alpha uv/(1-w)$	$2uv/(1-tw)-2\alpha uv/(1-w)$
$AA \times aa$	$\alpha uw/(1-w)+\beta u$	$uw(2-t)/(1-tw)-\alpha uw/(1-w)-\beta u$
$Aa \times Aa$	$\alpha v^2/(1-w)$	$v^2/(1-tw)-\alpha v^2/(1-w)$
$Aa \times aa$	$\alpha vw/(1-w)+\beta v$	$vw(2-t)/(1-tw)-\alpha vw/(1-w)-\beta v$
$aa \times aa$	βw	$w^2(1-t)/(1-tw)-\beta w$

(ii) Fertilities of matings

Mating	Fertility
$AA \times AA$	$[fu^2/(1-tw)+\alpha(1-f)u^2/(1-w)]/T$
$AA \times Aa$	$[2fuv/(1-tw)+2\alpha(1-f)uv/(1-w)]/T$
$AA \times aa$	$[fuw(2-t)/(1-tw)+\alpha(1-f)uw/(1-w)+\beta(1-f)u]/T$
$Aa \times Aa$	$[fv^2/(1-tw)+\alpha(1-f)v^2/(1-w)]/T$
$Aa \times aa$	$[fvw(2-t)/(1-tw)+\alpha(1-f)vw/(1-w)+\beta(1-f)v]/T$
$aa \times aa$	$[fw^2(1-t)/(1-tw)+\beta(1-f)w]/T$
$T = f+(1-f)(\alpha+\beta)$	

into selection against the dominant. If natural selection acts only on males, females mate with unchanged frequencies u, v and w.

After natural selection of males, matings then take place with female preferences α for the dominant A or β for the recessive a.

The frequencies of the components of the matings and their overall fertilities are shown in table 9.5. From this table the general recurrence equations of genotype frequencies can be obtained for models with natural selection and preferential matings of both phenotypes (see O'Donald 1980b). But this would be unnecessarily general to be applied to sexual and natural selection of the Arctic Skua.

The appropriate model for the Arctic Skua is a sexually favoured dominant which has a natural selective disadvantage as a result of later maturity. For this model $\beta = 0$, since pale males are not the objects of preferential mating. There are two cases of the model to consider.

Case (i). The dominant, preferred phenotype occurs at frequency $(1-w)/(1-tw)$. In this case all females can mate preferentially as a result of the condition

$$\alpha \leq (1-w)/(1-tw)$$

From table 9.5 we obtain the recurrence equations of the frequencies of genotypes in the next generation

$$Tu' = fp^2/(1-tw)+\alpha(1-f)p^2/(1-w)$$
$$Tv' = f(2pq-twp)/(1-tw)+\alpha(1-f)(2pq-pw)/(1-w)$$
$$Tw' = f(q^2-twq)/(1-tw)+\alpha(1-f)(q^2-qw)/(1-w)$$

giving

$$T\Delta p = \tfrac{1}{2}pw[ft(1-w)+\alpha(1-f)(1-tw)]/[(1-w)(1-tw)]$$

Since A is now the deleterious phenotype as a result of natural selection, we may put

$$t = -s/(1-s)$$
$$1-tw = (1-s+sw)/(1-s)$$

to obtain the recurrence equation

$$T\Delta p = \tfrac{1}{2}pw[-fs(1-w)+\alpha(1-f)-\alpha s(1-f)(1-w)]/[(1-w)$$
$$(1-s+sw)]$$

Thus A reaches equilibrium at the frequency

$$1-w_* = \frac{\alpha(1-f)}{s(\alpha+f-\alpha f)}$$

The difference equation for the gene frequency may be given in terms of the equilibrium w_* as follows:

$$T\Delta p = \tfrac{1}{2}spw(w-w_*)(\alpha+f-\alpha f)/[(1-w)(1-s+sw)]$$

This form of the difference equation shows that the frequencies globally converge on the point w_*. When $w > w_*$, $\Delta p > 0$ and $w' < w$; when $w < w_*$, $\Delta p > 0$ and $w' > w$. Unlike the models which allow only for sexual selection, these models, with monogamy, give different equilibria to those with polygyny. In a model with polygynous sexual selection, an equilibrium is reached at the point

$$1-w_* = \alpha/s$$

In monogamous sexual selection, when balanced by natural selection, the fertility parameter f now partly determines the position of equilibrium. This is what we should expect. Sexual selection is much less intense in monogamous species. Its intensity depends on f. Natural selection, being independent of the mating system, will have a much greater effect in monogamous species; the point of equilibrium will depend on the relative intensities of natural and sexual selection and hence on f.

Sexual selection of a deleterious recessive produces a corresponding equilibrium to sexual selection of a deleterious dominant. If β is the preference for the recessive, with selective coefficient t, then its equilibrium is reached at the frequency

$$w_* = \frac{\beta(1-f)}{t(\beta+f-\beta f)}$$

This corresponds exactly to the expression for the equilibrium frequency of a dominant.

Case (ii). The females with preferences are now assumed to outnumber the males they prefer:

$$\alpha > (1-w)/(1-tw)$$

This condition would apply when the sexually favoured dominant has recently entered the population and is increasing in frequency by sexual selection. Then we obtain the recurrence equations

$$Tu' = p^2/(1-tw)$$
$$Tv' = (2pq-pw)/(1-tw)+fpw(1-t)/(1-tw)$$
$$Tw' = (q^2-qw)/(1-tw)+fqw(1-t)/(1-tw)$$

and hence

$$T\Delta p = \tfrac{1}{2}pw(1-f+ft)/(1-tw)$$

Putting $t = -s/(1-s)$ leads to the equation

$$T\Delta p = \tfrac{1}{2}pw(1-s-f)/(1-s+sw)$$

This shows that if $s < 1-f$, $\Delta p > 0$ and hence apparently $p_n \to 1.0$. However, as p_n increases, the point must be reached at which

$$\alpha \leq (1-s)(1-w)/(1-s+sw)$$

and the condition for Case (ii) becomes that for Case (i) giving the equilibrium

$$1-w_* = \frac{\alpha(1-f)}{s(\alpha+f-\alpha f)}$$

If $s > 1-f$, $p_n \to 0$ and A is eliminated: Case (ii) continues to hold.

In general we can conclude that provided the sexually favoured phenotype has an initial advantage it can start to spread through the population until it reaches its equilibrium frequency $1-w_*$. The expression for $1-w_*$ can exceed the value 1.0 depending upon the values of α, s and f. In general when s and f are small, the value of $1-w_*$ will often exceed 1.0. Then, of course, A reaches a state of fixation at $1-w=1.0$.

So far we have considered the case when natural selection acts, like sexual selection, only on males. The females carry the genotypes AA, Aa

and *aa* with frequencies u, v and w but are not naturally selected as a result of carrying them. This model would apply to a sex-limited male character which is not expressed in females – a display character like the red throat of the stickleback or the tail of the peacock for example. But in the Arctic Skua, both sexes show the polymorphism. Even though the females are not sexually selected, the evidence suggests that the earlier onset of maturity of pale birds occurs in both sexes. Dark and intermediate birds will suffer a disadvantage in both sexes as a result. We now therefore analyse a model in which both sexes are naturally selected and males alone are sexually selected. In this model males and females are selected differently. This produces a state of equilibrium at which the Hardy-Weinberg Law no longer holds, as it always does when only one sex, usually the males, is selected. When both males and females are naturally selected, they both mate with the phenotypes at frequencies $(1-w)/(1-tw)$ and $w(1-t)/(1-tw)$ in the population. Table 9.6 shows the fertilities of the matings that are produced by preferential and random mate selection. Two cases must be considered when a dominant phenotype is the object of female preference ($\beta = 0$):

Case (i), in which $\alpha \leq (1-w)/(1-tw)$; and

Case (ii), in which $\alpha > (1-w)/(1-tw)$.

Case (i). This case, with condition $\alpha \leq (1-w)/1-tw)$, holds if the dominant phenotype has become not uncommon in the population. From table 9.6 we get recurrence equations

$$Tu' = fp^2/(1-tw)^2 + \alpha(1-f)p^2/[(1-w)\ (1-tw)]$$
$$Tv' = f(2pq-twp)/(1-tw)^2 + \alpha(1-f)\ (2pq-pw-twp)/$$
$$[(1-w)\ (1-tw)]$$
$$Tw' = f(q-tw)^2/(1-tw)^2 + \alpha(1-f)\ (q-w)\ (q-tw)/$$
$$[(1-w)\ (1-tw)]$$

where

$$T = f + \alpha(1-f)$$

Substituting $-s/(1-s)$ for t, where s measures the selective disadvantage of the dominant, we obtain the difference equation

$$T\Delta p = pw[\tfrac{1}{2}\alpha(1-f)-s(\alpha+f-\alpha f)\ (1-w)]/[(1-w)\ (1-s+sw)]$$

so that at equilibrium A is found to occur at the frequency

$$1-w_* = \frac{\tfrac{1}{2}\alpha(1-f)}{s(\alpha+f-\alpha f)}$$

Table 9.6. *Fertilities of the matings with preferences for both dominant and recessive phenotypes*

Mating	Fertility
$AA \times AA$	$\{fu^2/(1-tw)^2 + \alpha(1-f)u^2/[(1-w)(1-tw)]\}/T$
$AA \times Aa$	$\{2fuv/(1-tw)^2 + 2\alpha(1-f)uv/[(1-w)(1-tw)]\}/T$
$Aa \times Aa$	$\{fv^2/(1-tw)^2 + \alpha(1-f)v^2/[(1-w)(1-tw)]\}/T$
$AA \times aa$	$\{2fuw(1-t)/(1-tw)^2 + \alpha(1-f)(1-t)uw/[(1-w)$ $(1-tw)] + \beta(1-f)u/(1-tw)\}/T$
$Aa \times aa$	$\{2fvw(1-t)/(1-tw)^2 + \alpha(1-f)(1-t)vw/[(1-w)$ $(1-tw)] + \beta(1-f)v/(1-tw)\}/T$
$aa \times aa$	$\{fw^2(1-t)^2/(1-tw)^2 + \beta(1-f)(1-t)w/(1-tw)\}/T$ $T = f + (1-f)(\alpha+\beta)$

Given the same values of the parameters, this is the same phenotypic frequency as that for the recessive. It is easily shown to be globally stable.

Case (ii). This case, with condition $\alpha > (1-w)/(1-tw)$, must hold if A is a rare allele. Consequently, we have recurrence equations

$$Tu' = p^2/(1-tw)^2$$
$$Tv' = [2pq - 2twp - pw(1-f)(1-t)]/(1-tw)^2$$

and

$$Tp' = p[1 - tw - \tfrac{1}{2}w(1-f)(1-t)]/(1-tw)^2$$

where

$$T = f + (1-f)(1-w)/(1-tw)$$
$$= 1 - (1-f)(1-t)w/(1-tw)$$

Then we obtain the difference equation

$$T\Delta p = pw\{t + \tfrac{1}{2}(1-f)(1-t) - tw[t + (1-f)(1-t)]\}/(1-tw)^2$$

giving the equilibrium

$$\tilde{w}_* = \frac{t + \tfrac{1}{2}(1-f)(1-t)}{t[t + (1-f)(1-t)]}$$

Since this represents the case when the dominant is deleterious, we again substitute $-s/(1-s)$ for t and thus obtain

$$1 - \tilde{w}_* = \frac{fs - \tfrac{1}{2}(1-f)(1-s)}{s[fs - (1-f)(1-s)]}$$

which is the same expression as that for the recessive when $s = t$. This equilibrium represents the point

$$\alpha = (1-w)/(1-tw)$$

and hence the point at which Case (ii) changes into Case (i). From this point, therefore, the population moves to the equilibrium at which

$$1-w_* = \frac{\frac{1}{2}\alpha(1-f)}{s(\alpha+f-\alpha f)}$$

In the analyses of these models of monogamous sexual selection, we have seen that when sexual selection alone acts on the males of a population, the evolutionary outcome is the same in both polygynous and monogamous species. The final equilibrium is independent of that reduction in the fertility of later matings that is necessary for sexual selection to occur in a monogamous species. Any reduction in fertility, however slight, is sufficient to take the population to the same equilibrium. Only when all matings are equally fertile does the system of preferential and random mating become equivalent to random mating of all genotypes with no selection taking place. But the sexual selection can be very slow if the reduction in fertility is slight. Polygyny always produces much faster sexual selection.

With the advent of natural selection, the equilibria produced by monogamous and polygynous sexual selection are no longer the same. The reduction in the fertility of later matings partly determines the equilibrium point in monogamous species. If the preferred males gain an initial sexual advantage that outweighs the disadvantage of natural selection, the preferred phenotype will start to increase in frequency. It will either reach a polymorphic equilibrium or go to complete fixation.

The general computer model, which simulates the details of mate selection through the successive periods of the Arctic Skua's breeding season, produces exactly the same final equilibrium as the simplified models of the breeding season analysed in this section. (O'Donald, 1973, 1974). The results we have obtained can be applied in general to sexual selection in monogamous species, including the Arctic Skua. The models can all produce stable polymorphisms. For a given set of parameter values, only one point of equilibrium exists: all frequencies are attracted to this point; convergence is global. Given estimates of the parameter values – the mating preferences and the coefficients of natural selection – a unique outcome of the evolutionary process can be predicted. Models for the estimation of mating preferences can be tested by their goodness of fit to the frequencies of matings. They can also be tested independently by comparing the equilibrium in the model with equilibrium of the population.

9.3 Genetic models of male competition

Males often compete between themselves for mates: they may fight, threaten and defend territories. They may also be said to compete by trying to outdo one another in courtship and sexual display, but this competition implies that ultimately the females choose, depending on what the males have offered. In a competition by courtship, the female referees the contest and decides who was the winner. When fighting, threatening or defending territories, the combatants alone decide the outcome. The female stays on the sideline awaiting the victor. This difference produces a profound difference in the selection of the males. When females are in a position to choose, the selective advantage is always to some extent frequency-dependent: the advantage depends on the ratio of particular phenotypes of males to the females who prefer them; rare males gain a relatively greater advantage, since a lot of females prefer them, than common males who are in less demand. In polygynous species, rare males mate more often when they are rare; in monogamous species, they mate earlier in the breeding season.

When males fight or compete among themselves for females, the better competitors gain no inevitable frequency-dependent advantage. If one phenotype of male is better armed than another, the relative frequencies of the victories of the one over the other will in no way depend on their relative frequencies in the population. It will depend only on the relative advantage gained from possessing more effective spurs, claws, beak etc. for fighting. The advantage gained from defending a larger territory will not be frequency-dependent if females exercise no choice in favour of males with larger territories. If females simply land at random on the breeding grounds, the holders of larger territories always gain the same relative advantage over the holders of smaller territories, regardless of the relative frequencies of the larger and smaller territories. Direct male competition could only be frequency-dependent if the males varied the intensity of their competition in relation to frequency – a very implausible hypothesis.

Charlesworth & Charlesworth (1975) analysed a general model of male competition containing no element of female choice. They simply assumed that one phenotype of male was k times more likely to succeed in finding a mate than another phenotype. They added the complication that competition took place in groups of n males. In each group, the frequencies of the phenotypes were binomially distributed. Unfortunately this complication, which is biologically very reasonable, makes a mathematical analysis of the model impossible. The Charlesworths resorted to numerical computation. O'Donald (1980*a*) gave an analysis of a simpler version of the Charlesworths' model. In this version, the males with phenotypes *A* and *a* mate polygynously in the ratio $k:1$. Since these phenotypes exist in the population at frequencies $1-w$ and w, the males mate with probabilities

$$\frac{k(1-w)}{k(1-w)+w} \text{ and } \frac{w}{k(1-w)+w}$$

In this model, polymorphisms could be maintained only if the heterozygous males were the best competitors. In all other genetical systems, the constant advantage of the one phenotype over the other ensures fixation of the advantageous type.

The Charlesworths' model is implicitly based on the assumption of polygynous mating: some males mate k times more often than others. A model based on this idea can be constructed for a monogamous species if we assume that one phenotype is k times more likely to mate with an early female than with a late female. For example, if *A* is k times more likely to mate early than *a*, then among early matings, *A* and *a* will mate with probabilities

$$k/(k+1) \text{ and } 1/(k+1)$$

while in late matings they will mate with probabilities

$$1/(k+1) \text{ and } k/(k+1)$$

This reversal in the ratios must occur because the overall probabilities of monogamous matings must be equal.

9.3.1 *Competition between dominant and recessive phenotypes*

The *A* phenotype is genotypically either *AA* or *Aa*. Both genotypes are k times more likely to mate with early females than the recessive *aa*. Table 9.7 shows the relative frequencies of the early and late matings and their average fertilities. As in the models of the previous section, the fertilities are assumed to be in the ratio $1:f(f<1)$.

The matings and their fertilities give the following genotypic frequencies in the next generation:

$$Tu' = p^2(k+f)$$
$$Tv' = (2pq-pw)(k+f)+pw(kf+1)$$
$$Tw' = (q^2-qw)(k+f)+qw(kf+1)$$

where

$$T = (1-w)(k+f)+w(kf+1)$$

Table 9.7. *Frequencies of early and late matings and fertilities of matings determined by competition between dominant and recessive male phenotypes*

(i) Frequencies of early matings and late matings

Mating	Frequency of early matings	Frequency of late matings
$AA \times AA$	$ku^2/(k+1)$	$u^2/(k+1)$
$AA \times Aa$	$2kuv/(k+1)$	$2uv/(k+1)$
$AA \times aa$	$(k+1)uw/(k+1)$	$(k+1)uw/(k+1)$
$Aa \times Aa$	$kv^2/(k+1)$	$v^2/(k+1)$
$Aa \times aa$	$(k+1)vw/(k+1)$	$(k+1)vw/(k+1)$
$aa \times aa$	$w^2/(k+1)$	$kw^2/(k+1)$

(ii) Fertilities of matings

Mating	Fertility
$AA \times AA$	$[u^2(k+f)/(k+1)]/T$
$AA \times Aa$	$[2uv(k+f)/(k+1)]/T$
$AA \times aa$	$[uw(k+1)\,(f+1)/(k+1)]/T$
$Aa \times Aa$	$[v^2(k+f)/(k+1)]/T$
$Aa \times aa$	$[vw(k+1)\,(f+1)/(k+1)]/T$
$aa \times aa$	$[w^2(kf+1)/(k+1)]/T$
$T = [(1-w)\,(k+f)+w(kf+1)]/(k+1)$	

Then

$$T\Delta p = \tfrac{1}{2}pw\,(k-1)\,(1-f)$$

so that $\Delta p > 0$ for all $k > 1$, $f < 1$ which are the premises of the model of the Arctic Skua. The better competitors with phenotype A eliminate the others with phenotype a: $p_n \to 1.0$.

The results of male competition in monogamous and polygynous species are ultimately the same. The best competitor replaces the others. The difference equations for the two mating systems are very similar.

With monogamy: $\Delta p = \dfrac{\tfrac{1}{2}pw(k-1)\,(1-f)}{k+f-w(k-1)\,(1-f)}$

With polygyny: $\Delta p = \dfrac{\tfrac{1}{2}pw(k-1)}{k-w(k-1)}$

Polygyny is faster except when $f = 0$. At $f = 0$, the models are identical.

9.3.2. Competition between each genotype
This will occur if each genotype fares differently in competition. In

Table 9.8. *Frequencies of early and late matings and fertilities of matings determined by competition between male genotypes*

(i) Frequencies of early matings

Females	Males		
	AA	Aa	aa
AA	$ku^2/(k+l+1)$	$luv/(k+l+1)$	$uw/(k+l+1)$
Aa	$kuv/(k+l+1)$	$lv^2/(k+l+1)$	$vw/(k+l+1)$
aa	$kuw/(k+l+1)$	$lvw/(k+l+1)$	$w^2/(k+l+1)$

(ii) Frequencies of late matings

Females	Males		
	AA	Aa	aa
AA	$(l+1)u^2/(k+l+1)$	$(k+1)uv/(k+l+1)$	$(k+l)uw/(k+l+1)$
Aa	$(l+1)uv/(k+l+1)$	$(k+1)v^2/(k+l+1)$	$(k+l)vw/(k+l+1)$
aa	$(l+1)uw/(k+l+1)$	$(k+1)vw/(k+l+1)$	$(k+l)w^2/(k+l+1)$

(iii) Fertilities of matings

Mating	Fertility
$AA \times AA$	$[(k+lf+f)u^2]/T$
$AA \times Aa$	$[(k+lf+f+l+kf+f)uv]/T$
$AA \times aa$	$[(k+lf+f+kf+lf+1)uw]/T$
$Aa \times Aa$	$[(l+kf+f)v^2]/T$
$Aa \times aa$	$[(l+kf+f+kf+lf+1)vw]/T$
$aa \times aa$	$[(kf+lf+1)w^2]/T$
	$T = u(k+lf+f)+v(l+kf+f)+w(kf+lf+1)$

the simplest model males AA, Aa and aa mate with probabilities

$$k/(k+l+1),\ l/(k+l+1) \text{ and } 1/(k+l+1)$$

Table 9.8 gives the matings and their fertilities. Then the frequencies of the genotypes are given by the recurrence equations

$$Tu' = up(k+lf+f)+\tfrac{1}{2}vp(kf+l+f)$$
$$Tv' = uq(k+lf+f)+\tfrac{1}{2}v(kf+l+f)+wp(kf+lf+1)$$
$$Tw' = \tfrac{1}{2}vq(kf+l+f)+wq(kf+lf+1)$$

Hence we obtain the difference equation

$$T\Delta p = \tfrac{1}{2}\{u(k+lf+f)+\tfrac{1}{2}v(kf+l+f)$$
$$-p[u(k+lf+f)+v(kf+l+f)+w(kf+lf+1)]\}$$

Thus at equilibrium we see that

$$p = \frac{u(k+lf+f)+\tfrac{1}{2}v(kf+l+f)}{u(k+lf+f)+v(kf+l+f)+w(kf+lf+1)}$$

Evidently the Hardy-Weinberg Law holds at this point. Thus we can put p^2 for u, $2pq$ for v and q^2 for w. The expression then becomes quadratic in p, giving the solution

$$p_* = (l-1)/(2l-k-1)$$

If we assume that the Hardy-Weinberg Law also holds near equilibrium, then the difference equation can be written in the form

$$T\Delta p = \tfrac{1}{2}pq(1-f)[p(k-2l+1)-(1-l)]$$
$$= \tfrac{1}{2}pq(p-p_*)(k-2l+1)$$

The stability of the equilibrium point p_* depends on the sign of the expression $k-2l+1$. Suppose $l<\tfrac{1}{2}(k+1)$. The expression is then positive. If $p>p_*$, $\Delta p>0$ and $p'>p$: p increases and continues to do so until $p = 1.0$. If $p<p_*$, $\Delta p<0$ and p decreases until $p = 0$. The conditions $p>p_*$ and $p<p_*$ define domains of attraction to the fixation states $p = 1.0$ and $p = 0$. But if $l>\tfrac{1}{2}(k+1)$, the expression $k-2l+1$ is negative. Now p always converges to p_*. In these circumstances we say that p_* is an unstable equilibrium point if $l<\tfrac{1}{2}(k+1)$ and a stable point if $l>\tfrac{1}{2}(k+1)$. The equilibrium is only stable if the heterozygotes are the best competitors. Natural selection only produces a stable polymorphism with constant selective coefficients if the heterozygotes have the advantage. Similarly, male competition only does so if the heterozygotes are the best competitors.

These results are closely paralleled by those when the males compete polygynously. Then the difference equation becomes

$$\Delta p = \tfrac{1}{2}pq[p(k-2l+1)-(1-l)]/(ku+lv+w)$$

producing the same equilibrium with the same stability conditions as monogamous competition.

9.3.3 *Natural selection with competition between dominant and recessive phenotypes*

Since male competition does not produce frequency-dependent selection, we should not expect that it would maintain polymorphisms except when heterozygous males are the best competitors. We have seen that this expectation is true for the models of male competition we have analysed in the previous sections. For the same reason, we should not expect that natural selection would balance sexual selection by male competition at a point of polymorphic equilibrium. Either natural selection will outweigh sexual selection or sexual selection will outweigh natural selection. A character sexually selected as a result of male competition will either spread through a population to complete fixation or never spread at all.

Table 9.9. *Fertilities of matings with natural selection and competition between dominant and recessive phenotypes*

Mating	Fertility
$AA \times AA$	$\{u^2(k+f)/[(k+1)\,(1-tw)]\}/T$
$AA \times Aa$	$\{2uv(k+f)/[(k+1)\,(1-tw)]\}/T$
$AA \times aa$	$\{uw[(k+1)\,(f+1)-t(kf+1)]/[(k+1)\,(1-tw)]\}/T$
$Aa \times Aa$	$\{v^2(k+f)/[(k+1)\,(1-tw)]\}/T$
$Aa \times aa$	$\{vw[(k+1)\,(f+1)-t(kf+1)]/[(k+1)\,(1-tw)]\}/T$
$aa \times aa$	$\{w^2(1-t)\,(kf+1)/[(k+1)\,(1-tw)]\}/T$

$$T = \frac{(k+f)\,(1-w)+w(kf+1)\,(1-t)}{(k+1)\,(1-tw)}$$

Suppose the dominant phenotype increases competitive ability but lowers the chance of survival. Before competing for mates, the male phenotypes A and a have frequencies

$$(1-w)/(1-tw) \quad \text{and} \quad w(1-t)/(1-tw)$$

where $t = -s/(1-s)$ and measures the selective disadvantage of A. The fertilities of the matings are shown in table 9.9.

Putting $-s/(1-s)$ for t in the recurrence equations

$$Tu' = p^2(k+f)/[(k+1)\,(1-tw)]$$
$$Tv' = [(2pq-pw)\,(k+f)+pw(kf+1)\,(1-t)]/[(k+1)\,(1-tw)]$$

we obtain

$$\Delta p = \frac{\tfrac{1}{2}pw[(k-1)\,(1-f)-s(k+f)]}{[k+f-w(k+f)](1-s)+w(kf+1)}$$

This shows that A fixes if

$$s < \frac{(k-1)\,(1-f)}{k+f}$$

and a fixes if

$$s > \frac{(k-1)\,(1-f)}{k+f}$$

These results confirm what we expected. If $f = 1$ or $k = 1$, a always fixes since A has no competitive superiority. If $f = 0$, a fixes if

$$s > (k-1)/k$$

This, too, we should expect. It is also the result obtained if males compete polygynously. The relative selective advantage from competition is then

$$k:1$$

or $1:k^{-1}$

with competitive selective coefficient in favour of A given by the expression

$$1-k^{-1} = (k-1)/k$$

Since s is the natural selective coefficient in favour of a, we should then expect that A would fix if

$$s < (k-1)k$$

and a would fix if

$$s > (k-1)/k$$

Exactly analogous results to these can be derived for the case when a recessive is the best competitor.

We conclude from these analyses that male competition can only give rise to stable polymorphisms if the heterozygous males are the best competitors. In all other cases, one or other homozygous genotype becomes fixed in the population.

9.4 Preferential mating with assortment

In all the models analysed in previous sections, no assortative mating has been assumed. If females expressed preferences for particular male phenotypes, they did so whether or not they themselves possessed the phenotype they preferred in the males. In the simple case of two phenotypes A and a (where A has always been considered dominant to a), females who possessed either of these phenotypes (or who carried the genotypes that determine their expression in the males) would each mate preferentially in the same proportions. If α and β are the proportions mating preferentially with A and a males, then the frequencies of preferential matings would be

		Males	
		A	a
Females	A	$\alpha(1-w)$	$\beta(1-w)$
	a	αw	βw

This formulation leads directly to the model of preferential mating with dominance analysed in 9.2.1. But preferences might have been expressed only when females possessed the same phenotypes or genotypes as the males they preferred. Then the preferential matings would be

	Males	
	A	a
Females A	$\alpha(1-w)$	—
a	—	βw

The A females would thus tend to assort with the A males and the a females with the a males. This positive association or assortment of phenotypes in the matings is called *positive assortative mating*, or simply *assortative mating*. The assortment can also be of different, or unlike phenotypes, giving matings

	Males	
	A	a
Females A	—	$\beta(1-w)$
a	αw	—

This negative association in the matings is called *negative assortative* or *disassortative mating*.

Assortative mating must depend on inhibition of certain matings. In higher organisms, especially birds, this would usually act through the control of female sexual response. For example, in section 9.1 I explained the disassortative mating of melanic and non-melanic pigeons by females being inhibited from mating with the sexier melanic males if they were also melanic themselves. Davis & O'Donald's model of this hypothesis (Davis & O'Donald, 1976*b*) fits the data of mating frequencies extremely well. In the polymorphism of melanic and non-melanic forms of the Snow Goose (Cooke, 1978; Cooke & Cooch, 1968; Cooke, Mirsky & Seiger, 1972) evidence strongly supports the hypothesis that the assortative mating is determined by preferences acquired by imprinting of the parental phenotypes. Imprinting necessarily produces positive assortative mating (O'Donald, 1960*b*), since offspring then prefer to mate with phenotypes like their parents and hence more often with phenotypes like themselves.

Appendix C describes the maximum likelihood estimation of an assortative or disassortative mating preference. The estimates thus obtained are used in deriving tests of the goodness of fit of the models (appendix C, section C.3).

Positive assortative mating is likely to be much the more common form of assortment. It will arise whenever divergence is taking place between populations characterized by different phenotypes. In one area one phenotype may be common, while in another area, it may be rare. If each sub-population is adapted to its own area, hybrid mating will disrupt the co-adapted genomes. Assortative mating will thus tend to evolve as the sub-populations diverge. Positive assortative mating is indeed almost inevitable if female preferences have evolved by the selective advantage of mating with advantageous males. As it evolves, the preference adds to the initial advantage and Fisher's 'runaway process' results. But this process entails the evolution of some degree of positive assortment in the expression of the females' preference: a gene for the preference is selected precisely because the offspring of preferential matings tend to carry both the gene for the preferred character and the gene for the preference. Hence the preference tends to be expressed by females who carry the gene for the preferred character: preferential matings tend to become assortative. In *Genetic Models of Sexual Selection* (1980*a*), I analysed some genetic models of the evolution of mating preferences. In the course of evolution, 10–20 per cent of the preferential matings become assortative. Karlin & O'Donald (1978) and O'Donald (1980*a*) analysed general models for polygynous species with mixed assortative mating and sexual selection. Much earlier, O'Donald (1960*a*) had set up models of assortative mating for monogamous species. However, in these original models of assortative mating, it was assumed that all matings were equally fertile: assortative mating then affects only the genetic structure of the population; it produces deviations from the Hardy-Weinberg Law, but does not change the gene frequencies in the course of evolution.

In this section I shall consider the evolutionary consequences of assortative mating with dominance. I assume the females' preferences can be expressed either with or without assortment. The assortative preference is always expressed as a fraction of the preferred phenotype: the females with preferences can never exceed in number the males they prefer. The preferences are defined by the table:

	Males	
	A	a
Females A	$\alpha(1-w)$	—
a	—	βw

Within the A phenotypes, the genotypes AA and Aa mate randomly. This gives rise to the matings and fertilities shown in Table 9.10. Since monogamy prevails, no selection occurs unless some matings are less fertile than others. As in the non-assorting models of preferential mating, I assume that after females with preferences have mated, the remaining females mate at random with fertility $f(f<1)$. Thus the table shows the preferential matings taking place with fertility 1 and the random matings taking place with fertility f.

Table 9.10. *Fertilities of matings in a model of monogamous assortative mating*

Mating	Fertility
$AA \times AA$	$\dfrac{\alpha u^2}{1-w} + \dfrac{fu^2(1-\alpha)^2}{1-\alpha(1-w)-\beta w}$
$AA \times Aa$	$\dfrac{2\alpha uv}{1-w} + \dfrac{2fuv(1-\alpha)^2}{1-\alpha(1-w)-\beta w}$
$AA \times aa$	$\dfrac{2fuw(1-\alpha)\,(1-\beta)}{1-\alpha(1-w)-\beta w}$
$Aa \times Aa$	$\dfrac{\alpha v^2}{1-w} + \dfrac{fv^2(1-\alpha)^2}{1-\alpha(1-w)-\beta w}$
$Aa \times aa$	$\dfrac{2fvw(1-\alpha)\,(1-\beta)}{1-\alpha(1-w)-\beta w}$
$aa \times aa$	$\beta w + \dfrac{fw^2(1-\beta)^2}{1-\alpha(1-w)-\beta w}$

The total fertility of all matings is then

$$T = \alpha(1-w) + \beta w + f[1 - \alpha(1-w) - \beta w]$$
$$= f + \alpha(1-f)\,(1-w) + \beta w(1-f)$$

The recurrence equations of genotypic frequencies are therefore

$$Tu' = \frac{\alpha p^2}{1-w} + \frac{fp^2(1-\alpha)^2}{1-\alpha(1-w)-\beta w}$$

$$Tv' = \frac{\alpha pv}{1-w} + \frac{f(1-\alpha)[2pq - \alpha vp - 2\beta wp]}{1-\alpha(1-w)-\beta w}$$

giving the recurrence equation of gene frequency

$$Tp' = p + fp(1-\alpha)$$

and hence the difference equation

$$T\Delta p = pw(1-f)\,(\alpha-\beta)$$

This shows that, exactly as when polygynous species mate assortatively, either one or other allele has a consistent advantage and proceeds to fixation: if $\alpha > \beta$, $\Delta p > 0$ and A fixes; if $\alpha < \beta$ $\Delta p < 0$ and a fixes. No polymorphic equilibrium can exist.

A polymorphism can exist and be maintained, however, if both assorting and non-assorting preferences are expressed for the two phenotypes. The equilibrium frequencies are identical to those obtained by Karlin & O'Donald (1978) for a polygynous species. Then an equilibrium is established at the frequency

$$w_* = \frac{\zeta+\eta+\alpha-\beta-[(\zeta+\eta+\alpha-\beta)^2-4\eta(\alpha-\beta)]^{\frac{1}{2}}}{2(\alpha-\beta)}$$

where ζ and η are the non-assorting preferences for A and a respectively. O'Donald (1980a) fitted this general model to data of matings of the parasitic wasp, *Mormoniella vitripennis*. This was data of Grant, Snyder & Glessner (1974) who observed matings of mutant and wild-type males presented to females at different frequencies. Since the frequencies at which the males were presented to the females were fixed at the ratios wild-type (+):mutant (m) of

2 (+):8(m)
5 (+):5(m)
8 (+):2(m)

while equal numbers of wild-type and mutant females made their choices, a total of 12 classes of mating were observed giving nine degrees of freedom for the estimation of the four parameters of the model.

Unfortunately, this general model combining both assorting and non-assorting preferences cannot be fitted to data of Arctic Skua matings. With no assortment, mating preferences can only be estimated by observation of the sequence in which the matings take place during the breeding season. Over the whole of the season, the frequencies of the matings will be random. But if the phenotypes assort when they mate, the overall frequencies of the matings can be used to detect and estimate the assortative mating preferences. In principle, both assorting and non-assorting parameters could be estimated from data of the breeding dates of all different pairs of birds. But there are scarcely enough data to be classified into pairs of phenotypes, with nine different classes, rather than just the three different phenotypes. Assortative mating of dark and intermediate phenotypes has been observed by counting numbers of pairs in several different colonies of Arctic Skuas in Shetland. In some of the colonies, the birds were only classified as melanic or non-melanic. But the model can be fitted to these data by assuming that $\beta = 0$. The expected frequencies of the matings can be obtained from table 9.10 by putting $\beta = 0$ and $f = 1$. Then the preference for melanics, α, can be estimated and the goodness of fit of the model tested on the data of each colony.

Assortative mating can also take place between genotypes. To describe a general model of preferential mating, three assorting and three non-assorting preferences must be defined: these will be the preferences for AA, Aa and aa genotypes. Karlin & O'Donald (1978) analysed this model. They found that three polymorphic equilibria existed and could be stable under certain conditions. There was a central equilibrium point with $p_* = \frac{1}{2}$ and two asymmetric equilibria at points on either side of the central point. This model can only be applied to data of pairs of Arctic Skuas classified as dark, intermediate or pale. Such data are of course available for the colonies of Arctic Skuas on Fair Isle and Foula. But the results of fitting the model are not significant given only these data. The details of these analyses of the data are given in the next chapter, section 10.3.

Assortative mating has been observed between melanic and non-melanic phenotypes in a number of different birds. Positive assortative mating is characteristic of the Arctic Skua and the Snow Goose. Negative assortative mating is very strong between melanic and non-melanic pigeons. It has also been found in one colony of Eleonora's Falcon (Walter, 1979). But another colony did not show it (see table C7, appendix C, section C.4.2). Assortative mating is easy to detect as a deviation from random mating. Since non-assorting preferences give rise to the same overall frequencies as random mating, preferential mating without assortment, unlike assortative mating, cannot be observed simply by counting the total numbers of matings: it can be observed in mating choice experiments, or in the sequence of matings as in the breeding dates of pairs of Arctic Skuas. Since assortative mating is a special type of preferential mating with expression of preference limited to only certain genotypes or phenotypes, it may be expected to be generally rarer than non-assortative preferential mating. Preferential mating may therefore be much more common than those instances of assortative mating that have been observed.

Assortative mating must always be strong evidence for some form of female choice. When some males court the females more actively and secure a higher proportion of the matings early in the season, assortative mating of the sexier males can only arise if at least some females are inhibited from mating with males like themselves. It is difficult to believe that male behaviour alone could produce assortative mating: the males would have to compete more strongly for females like themselves. This is possible in principle, though all evidence suggests that males are undiscriminating in their choice of mates.

10

Mating preferences of the Arctic Skua

10.1 Models of preferential mating

In this chapter, I bring together and analyse the different lines of evidence on the mechanism of sexual selection in the Arctic Skua. The models to be fitted to the data have already been described in sections 9.2 and 9.4. They are all models of preferential mating, not male competition. As we proved in section 9.3, male competition cannot maintain a polymorphism without heterozygous advantage. Heterozygotes suffer a slight disadvantage compared to dark homozygotes (section 8.5). Since the polymorphism of the Arctic Skua is stable, models of male competition cannot plausibly be fitted to the data.

Males who are the objects of female preference will tend to be chosen as mates earlier in the breeding season. We may postulate that a certain proportion of females will prefer to mate with particular male phenotypes. If the males have already arrived on their territories or joined a 'club' of unmated males competing for the females' attention, then as groups of females come into breeding condition they will be able to choose males from among those who are still unmated. If we are to simulate this process of mate selection as it goes on through the breeding season, we must allow for the intervals of time which are required for courtship and mate selection. On average, as we saw in section 8.4.2, new pairs take an extra 7–8 days to breed compared to old, established pairs. Dark males take about six days, intermediates eight days and pales 11 days to find mates. Courtship and mating may be accomplished in shorter, but not longer, times than these. We may plausibly assume that about a week is required for new pairs to form. Having divided the breeding season into weekly intervals, we may then assume that each week represents a period within which a group of females select their mates. The division into weeks can only be arbitrary. For the computer simulations, I took that division which had already been used to illustrate the distribution of breeding dates of new pairs (table 8.1, section 8.4.1, chapter 8). The proportions of all pairs breeding each week are equivalent to the proportions of females choosing their mates. These females exercise their choices of mates from among the remaining pool of unmated males. As the breeding season advances, fewer males are left to choose from, and females have less opportunity to mate preferentially. The matings that actually took place in a particular week can be allocated to dark, intermediate and pale males in accordance with a model of the expression of female preference. The aim is thus to fit alternative possible models of the females' preferences to the distributions of breeding dates.

O'Donald (1976, 1980*a*) described four models of preferential mating in favour of dark and intermediate Arctic Skuas. In the simplest model, which I call *Model 1*, a proportion β of the females prefer to mate with any melanic male, either intermediate or dark. The remaining $1 - \beta$ of the females mate randomly with any of the available males. Females who prefer melanic males will also mate with non-melanic pale males if all the melanics have already mated. A computer simulates the sequence of matings. In each interval in which groups of females choose their mates, the preferential matings may either precede or follow the random matings, or take place simultaneously with them. In what I call the P models, preferential matings precede random matings: females with preferences choose first; the remaining females then choose between the remaining males at random. At some point, too many females may be chasing too few melanic males. Females who have not been able to mate according to their preference, join the random mating females and mate with the only available males, all of whom will of course be non-melanic. The sequence of events can be slightly more complicated in the R models, in which random matings precede preferential matings. After the random matings, females with preferences choose the melanics among those males who are left. But if too few melanics are left for them, those females who could not satisfy their preference finally choose a non-melanic male. In the S model both random and preferential matings occur simultaneously. Both P and R models are equivalent to the S model when the breeding season is divided into very short intervals of time. Effectively the females then come singly to choose a male, each female removing one male successively from the pool of unmated males. In this case, P, R and S models all give exactly the same estimates of the mating preferences needed to fit the distributions of the breeding dates. The size of the interval into which the breeding season is divided is indeed a parameter of the models. I chose weekly intervals

Table 10.1. *Breeding dates of melanic and non-melanic male Arctic Skuas in new pairs*

Breeding dates in weekly intervals	Number of males breeding			
	D	I	P	Total
10–16 June	2	4	1	7
17–23 June	15	45	4	64
24–30 June	30	67	16	113
1–7 July	17	48	16	81
8–14 July	5	25	15	45
15–21 July	3	10	5	18
Total no.	72	199	57	328

because a week is about the period necessary for pairs to form; independently, it also produces the best-fitting model. It has been used, therefore, as the appropriate period for fitting the P and R versions of Model 1 to the data.

In *Model 2*, the females with the preference mate always with dark males if they can. When all dark males have found mates, the females mate with intermediate males; they only mate with pale males if no darks or intermediates are left unmated. A proportion α of the females thus mate preferentially. The remaining $1 - \alpha$ of the females mate randomly. The preferential matings may take place before, after or simultaneously with the random matings giving the versions 2P, 2R and 2S of Model 2.

There are two groups of females in *Model 3* with different preferences. One group of females, representing a proportion α of all females, prefer to mate only with dark males. With no dark males left to choose from, they then mate at random among any remaining intermediate or pale males. The other group of females prefer melanic males, either dark or intermediate. These represent a proportion β of the females. They mate in exactly the same way as the females in Model 1 who prefer melanic males. Whether they choose a dark or intermediate melanic male depends only on the proportions of dark and intermediate males among the melanics they choose from. The remaining $1 - \alpha - \beta$ of the females mate at random. The sequence of preferential and random matings gives rise to versions 3P, 3R and 3S of Model 3. If $\alpha = 0$, Model 3 is obviously identical to Model 1. Melanism is then fully dominant in its effect on female preference. If $\beta = 0$, melanism is recessive in the model. Otherwise melanism is semi-dominant: some females are sufficiently stimulated by either melanic phenotype to respond to courtship; others need to be courted by the dark melanic phenotypes.

In *Model 4*, a proportion α of the females prefer darks, while a separate proportion β prefer intermediates. The remaining $1-\alpha-\beta$ of the females mate randomly. This is a special case of the more general model in which females have separate preferences for each genotype. Model 4P, 4R and 4S are the three alternative versions of this model.

Model 4P corresponds to the model of separate preferences for each genotype analysed in section 9.2.2 of the previous chapter. Model 3P, on the other hand, gives rise to the selection of a dominant or recessive as analysed in section 9.2.1. In the next section, the models are fitted to the data of the males' breeding dates.

10.2 Estimation of mating preferences

Table 8.1 of chapter 8 shows the distributions of the breeding dates of dark, intermediate and pale males in new pairs. The breeding season is divided into weekly intervals. The table gives the numbers of males breeding in each weekly interval. We have already shown by analysis of variance of means and individual values, that the melanics breed very significantly earlier than pales. But this significant difference is observed only when males have taken new mates: it is therefore the consequence of sexual selection.

The dates when individuals are considered to have bred are the dates when their first egg hatched. This is the one event that can be dated reliably (see section 4.2, chapter 4). The pairing, and hence the sexual selection, took place several weeks previously. However, since the incubation period varies little between individuals or phenotypes, hatching date is a good, relative measure of breeding date. For the purpose of fitting the models, the relative dates are all we need. The aim in fitting the models is to find those values of the parameters of the mating preferences giving the theoretical distributions that agree as closely as possible with the actual distributions of table 8.1.

How can we decide what gives this closest fit? The usual statistical procedure is to find those values of the parameters that maximize the likelihood. Suppose that particular event i occurs with theoretical probability p_i. a_i such events have been observed. We define for all $i=1, 2 \ldots, k$ different possible events, the likelihood

$$L = \prod_{i=1}^{k} p_i^{a_i}$$

and the log likelihood

$$\log_e L = \sum_{i=1}^{k} a_i \log_e p_i$$

The probabilities p_i are obtained from the hypothetical model to be fitted to the data. They are functions of the parameters – the mating preferences α and β. The numbers a_i are the numbers of each phenotype breeding in each week of the season – in other words the values in table 8.1. The likelihood represents, for a given hypothesis with given parameter values, the relative probability of obtaining the observed numbers a_i ($i = 1, 2 \ldots, k$) if the hypothesis is true. The Principle of Likelihood (Edwards, 1972) tells us to choose the hypothesis and parameter values that maximize the likelihood (or equivalently the log likelihood, which is usually easier to handle mathematically). Likelihood differs fundamentally from probability. Probability is the relative frequency of the different possible outcomes, assuming the hypothesis is true: a_i varies while p_i is fixed. But likelihood takes the outcome as fixed and considers the possible variations in p_i given the particular set of observations a_i ($i = 1, 2 \ldots, k$). The values of the parameters that maximize $\log_e L$ are called the Maximum Likelihood estimates (M.L. estimates). As well as conforming to the general likelihood principle of scientific inference, the M.L. estimates have other desirable properties that estimates in general should have. In particular, they have the smallest sampling variance and are therefore the most accurate estimates that can be obtained. They summarize the data completely, providing in most cases all the information about the parameters that the data can possibly give. They connect closely with the χ^2 test of goodness of fit. But they can be mathematically difficult to calculate: the problem is to find the maximum value of the function $\log_e L$.

In the models of preferential mating, there are nine parameters to estimate: two mating preferences; two independent frequencies of the three phenotypes; and five independent probabilities of breeding in each of the six intervals. Since no heterogeneity in breeding dates exists between new pairs in the two periods 1948–62 and 1973–79, the data have been combined for the estimation of parameters. The combined data are shown in table 10.1. The marginal totals for phenotypes and intervals are divided by the grand total to give the M.L. estimates of phenotypic frequencies and probabilities of breeding in each interval. These estimates are the same for all models. They are used in the computer simulations of the models to calculate the proportions of the phenotypes that we expect to breed in each interval. The mating preferences are found by trial and error. I start with arbitrary values of the parameters. I then simulate the expected distribu-

Table 10.2. *Maximum likelihood (M.L.) estimates of female preferences when the models of preferential mating are fitted to the data of table 10.1*

Model	M.L. estimates of female preferences $\hat{\alpha}$	$\hat{\beta}$	Log likelihood (to base e)	Support for Model 3P
1P	—	0.382	− 501.466	0.471
1R	—	0.194	− 503.292	2.297
1S	—	0.286	− 502.285	1.290
2P	0.063	—	− 505.277	4.282
2R	0.026	—	− 505.724	4.729
3P	0.039	0.344	− 500.995	0
3R	0.022	0.172	− 502.842	1.847
3S	0.026	0.260	− 501.836	0.841
4P	0.114	0.262	− 501.120	0.125
4R	0.057	0.132	− 503.066	2.071
4S	0.085	0.195	− 501.997	1.002

In *Models 2, 3 and 4*, α is the proportion of females preferring dark males. In *Models 1 and 3*, β is the proportion preferring dark or intermediate males indiscriminately, or, in *Model 4*, the proportion preferring intermediate males.

tions of breeding dates and calculate the likelihood. New values of the parameters are produced by the computer as random deviations from the old values, subject to the constraint

$$\alpha + \beta \leq 1$$

The likelihood is calculated again. This process is repeated 100 times by the computer and the values giving the highest likelihood chosen for the next step. The maximum size of the random deviations is then reduced by half and the whole process repeated a further 100 times, the maximum likelihood values chosen and the maximum step size of the random deviations halved again. The computer repeats these cycles until the maximum step size is less than 0.00001. In this way, the computer produces a random walk up the likelihood surface of the model, using steps of ever-decreasing size until it has found the maximum. The simulation can be checked by starting it over a range of different initial values, to make sure the same maximum is always attained.

The results of these calculations are shown in table 10.2. The P models all give higher log likelihoods at their respective maxima than the R models. As expected, the likelihoods and estimates of the S models are intermediate between the P and R models. For comparisons of the fit of the different models, the relative support is also shown. Support is the relative increase in log likelihood when one hypothesis is substituted for another less likely one (Edwards, 1972). Model 3P has the highest likelihood, followed closely by Model 4P and Model 1P. The differences in their log likelihoods is less than 1.0, which is by no means significant. The R models are about two units of log likelihood below the P models. Thus substituting Model 3P for one of the R models raises the log likelihood by about two units. The support for Model 3P is the amount by which the log likelihood is raised when Model 3P is substituted for the model in question. Where that support is greater than two units, the model in question can be rejected, for the two unit level of support corresponds roughly to the five per cent level of significance in a normally distributed variate. Table 10.2 shows that Model 2 is completely ruled out.

Except for Model 2, the total preference is roughly the same in each of the P, R and S models. For example, the quantity $\hat{\alpha} + \hat{\beta}$ in Model 3P is equal to $\hat{\beta}$ in Model 1P, and so on. The proportion of females expressing preference is much greater in the P models than in the R models: if random matings take place first, as in the R models, a considerable proportion of preferred males will already have mated before preferential matings occur. A smaller proportion of females need express preferences to select the same proportion of preferred males as in the P models.

In section 9.1 of the last chapter, I argued that if some females have evolved preferences for particular male phenotypes, we should expect that preferential matings would precede random matings: the females will evolve lower thresholds of response towards the males they prefer and thus mate preferentially before they mate randomly. On the other hand, if the males should differ in the intensity of their courtship, while individual females respond at the same average threshold towards all males, then random matings would tend to precede preferential matings: a responsive female would mate first with any male; less responsive females would mate later, mainly with the more actively courting males. Thus the greater support for the P models provides significant evidence in favour of female preference. Both behavioural mechanisms depend on the females making the choices. But if female preferences evolve as Fisher suggested, then females must evolve differences in response, regardless of whether the males may differ in the intensity of their courtship.

Imagine the females are offered the males as if males were different products that could be chosen. The females may differ in their sales resistance to the products they are offered. Or they may have the same resistance to all products, but one product is being sold by more active and enthusiastic salesmen than the others. In both cases the females are the arbiters: they decide. But in the first case, we should say they have a definite preference in favour of a particular product, which they might choose regardless of the salesmanship. And the difference between a female having a preference in this ordinary sense of the word, and having to be pressured by salesmanship into choosing one product rather than another is the difference which gives rise to the P and R models. If she wants a particular product, she takes it without regard to the salesmanship. If she does not particularly want any of the products, a lot of sales pressure may be necessary to get her to choose one. Reluctant females with no particular preference take longer to choose, the product they choose depending on the stamina and persistence of the salesman.

In Models 1, 3 and 4, the P versions of the models fit better than the R and S versions. But how can we discriminate between Models 1P, 3P and 4P, none of which differ significantly in their likelihoods? Models 1P and 3P are virtually equivalent. The extra parameter of the preference for dark males in Model 3 differs not significantly from zero. Model 1 is just a special case of Model 3. But the extra parameter is an essential part of Model 4. Without the preference for dark males, the model would produce identical distributions of breeding dates for both darks and pales since neither would be selected. But it is in these distributions that darks and pales differ most significantly: comparing the distributions of darks and pales, we find $\chi^2_4 = 14.43$, $P = 0.00604$. Both parameters are therefore essential to the fitting of Model 4. The analysis of the model of separate preferences for each genotype (section 9.2.2) shows that an equilibrium will be established at the relative frequency

$$p_* = (\alpha + \tfrac{1}{2}\beta)/(\alpha + \beta)$$
$$= 0.652$$

The same formula applies to both R and S models: the estimates of the preferences for these models all give the same equilibrium frequency. As shown in section 9.2.2, this is a point of stable equilibrium: given the preferences, the gene frequency will be maintained at this value. But it is, as we should expect from the demographic analysis of chapter 6, a considerably higher frequency than the actual frequency of the allele for dark. Data of matings on Fair Isle (table 10.5) show that when the population is at a Hardy-Weinberg equilibrium, $q^2 = 78.5/392$ and the allele for dark has an estimated frequency $p = 0.553$. Its hypothetical frequency determined by sexual selection will of course be reduced by the natural selection favouring pale (chapter 6, section 6.3).

Model 3P, which fits the data slightly better than Model 4P, would not give rise to a polymorphism at all unless natural selection for pale balanced the sexual selection for dark and intermediate. According to Model 3P, as the analysis of section 9.2.1 has shown, dark has the overall advantage at all frequencies and eventually fixes in the population. The sexual selection in Model 3 entails fixation, while in Model 4 it entails polymorphism. The advent of natural selection in favour of pale gives rise to polymorphism in both models. Thus, although existence of a stable polymorphism excludes the hypothesis of sexual selection by direct competition of male against male, it does not exclude specific models of female preference. Melanism is a semi-dominant character, and Model 3P is precisely a model of the expression of preference for a semi-dominant male phenotype. Model 4P, however, is a model of the expression of preference for distinct heterozygous and homozygous phenotypes. On grounds of the genetics of melanism, Model 3P is more realistic.

So far, we have only succeeded in showing that Model 3P is the best of the models: it has a higher likelihood than any of the others. But this is not sufficient evidence that it fits the data. It may fit the data better than any other model, but still not well enough. To test its goodness of fit we must show that after fitting no significant residual variation is left to be explained. This can be done by an analysis of χ^2 based on the data of table 8.1. Several components of variation may determine the numbers observed in the table: mating preferences, as we have seen, certainly do so; phenotypic proportions may vary between the periods 1948–62 and 1973–79; so also may the breeding dates. The variation caused by these components can be extracted from the total variation to leave a residual level of variation caused by effects of random sampling alone. If the model fits the data satisfactorily, the value of χ^2 corresponding to this residual variation should not be statistically significant. If it were significant, we should be forced to conclude that the model is inadequate to explain all the variation in the data.

I analysed the data in the form shown in table 10.3. The observations of the first two and last two weeks of the breeding season were lumped together to give sufficiently large expected numbers in each class for the calculation of χ^2. The data thus consist of two 3×4 contingency tables, with a total of 24 classes. The proportions of pairs in the four intervals of the breeding season are estimated by three independent values. The overall frequencies of the three phenotypes have two independent estimates. The total numbers in each period are necessary to calculate the expectations based only on the overall phenotypic frequencies and the overall proportions breeding in each interval. A total of seven independent degrees of

Table 10.3. *Data for analysis of χ^2*

Breeding dates	Numbers observed in period 1948–62			Numbers observed in period 1973–79			Proportion of all birds in interval
	D	I	P	D	I	P	
10–23 June	9	26	4	8	23	1	0.2165
24–30 June	16	19	6	14	48	10	0.3445
1–7 July	10	24	6	7	24	10	0.2470
8–21 July	4	17	10	4	18	10	0.1921
Total of each phenotype	39	86	26	33	113	31	

Proportion of D = 0.2195
Proportion of I = 0.6067
Proportion of P = 0.1738

freedom have thus been lost from the 24 before the model of preferential mating has been fitted or the effects of variation between periods removed.

For these 17 degrees of freedom, we obtain the value of χ^2

$$\chi_{17}^2 = 31.6851$$

This corresponds to a probability of

$$P = 0.0165$$

The total variation is significantly greater than that which random sampling alone would produce. Of course, the analysis of variance (table 8.2, section 8.4.1, chapter 8), which provides a much more accurate test of significance, has already demonstrated this.

In obtaining the value of χ_{17}^2, the expected proportions of each phenotype in each period and each date interval were calculated simply by assuming independence of phenotypes, periods and intervals and multiplying together the proportions shown in the table. If the model of preferential mating does fit the data, the value of χ^2 will be reduced significantly after fitting, leaving no significant heterogeneity in the data. Using the M.L. estimates of parameters for Model 3P, the expected proportions for each of the 24 classes were then computed for this model and χ^2 calculated again. This gave the value

$$\chi_{15}^2 = 17.2417$$

corresponding to

$$P = 0.305$$

The model thus fits the data. But this χ_{15}^2 still contains contributions from

Table 10.4. *Analysis of χ^2 when Model 3P is fitted to the data of table 10.3*

Component of variation	Value of χ^2	Dfs	Value of P
Preferential mating of melanics	14.4434	2	0.000730
Differences in breeding dates			
between periods 1948–62 and 1973–79	7.2071	3	0.0656
Differences in phenotypic frequencies			
between periods 1948–62 and 1973–79	2.5570	2	0.278
Residual heterogeneity	7.4776	10	0.680
Total	31.6851	17	0.0165

variation not concerning the model of sexual selection – variation between periods in breeding dates and proportions of phenotypes. The numbers of phenotypes in each period form a 2×3 contingency table for which

$$\chi_2^2 = 2.5570$$

The proportion breeding in each interval form a 2×4 contingency table for which

$$\chi_3^2 = 7.2071$$

Thus, finally, we can set out a complete analysis of χ^2 as shown in table 10.4.

A very similar analysis is obtained when Model 4P is fitted to the data. Model 4P fits the data just as well as 3P. Analysis of χ^2 provides no means of discriminating between Models 3 and 4. Model 3 is to be preferred only on the *a priori* grounds that it represents preferential mating of a semi-dominant character such as the melanism of the Arctic Skua has been shown genetically to be (chapter 5).

The hypothetical distributions of breeding dates in Model 3P are shown in figure 10.1. These are the proportions of individual males breeding in each week. They were used to calculate the value of χ^2 for the fitted model. A much higher proportion of pale males are left unmated until the last week of the breeding season compared with intermediate and dark males. Relative to their frequency, the dark males always have the highest probability of finding a mate. Very few darks are left unmated at the start of the last week. They have the smallest variance in breeding date; whereas the pales, with a highly skew distribution of breeding dates, have the largest variance. The hypothetical variances (given in units of days^2) are as follows:

var (D) = 58.2
var (I) = 63.8
var (P) = 78.7

All these differences in the distributions of breeding dates disappear in pairs of birds breeding for two or more years together.

10.3 Assortative mating

I have argued that assortative mating will almost always depend on some form of female choice (see sections 9.1 and 9.4, chapter 9). Even if males differ in competitive ability, some mechanism must exist to inhibit a female from responding to certain males – to males like herself if assortment is negative or to males unlike herself if assortment is positive. In such a case, the preference would depend on a more ready response to the sexier males, but its assortative expression would depend on female inhibitions. As we have seen (chapter 8, section 8.2), the evolution of preference necessarily entails the evolution of some degree of assortative expression of preference since genes for the preference must become associated with genes for the preferred character (O'Donald, 1980a). But the assortative matings form only a small proportion (no more than 20 per cent) of all preferential matings. The estimates of preferences given in the previous section were obtained from the breeding dates of males mating with new females. These estimates include both assorting and non-assorting prefer-

Figure 10.1 Theoretical distributions of the breeding dates of the phenotypes.

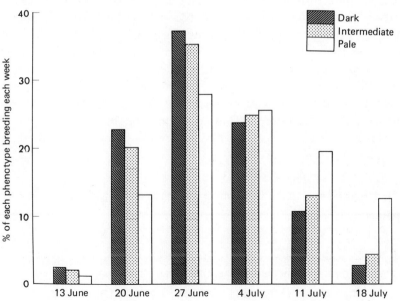

Table 10.5. *Matings between phenotypes of Arctic Skuas on Fair Isle 1973–79*

Mating	Years							
	1973	1974	1975	1976	1977	1978	1979	Total
D × D	4	3	4	4	1	1	2	19
D × I	25	7	12	11	16	10	13	94
D × P	11	4	5	2	3	4	10	39
I × I	40	23	21	9	16	13	19	141
I × P	19	9	8	17	13	7	7	80
P × P	5	5	3	1	2	3	0	19
Total	104	51	53	44	51	38	51	392

In this table all matings in 1973 and new matings in subsequent years are included.

ences. Since there are insufficient data to analyse the matings of the males in terms of the phenotypes of the females they mated with, the assorting and non-assorting components of the preferences cannot be separated.

But the overall frequency of each different type of mating is also known. As shown in appendix C, section C.2, these frequencies can be used to estimate the assorting components of the preferences. In 1975, matings of Arctic Skuas on the islands of Fair Isle and Foula showed assortative mating of pale and intermediate birds (Davis & O'Donald, 1976a). But subsequent data that included all matings on Fair Isle in the period 1973–79 gave no significant evidence of assortment, though intermediate × intermediate matings were in excess of the expectations of random mating. 1975 seems to have been a freak year in the number of these matings. The data of all new pairs formed since 1973 are shown in table 10.5.

This is a 7×6 contingency table. It shows no significant variation in frequencies between years:

$$\chi^2_{30} = 36.484, \quad P = 0.193$$

The total numbers of the six mating types show no deviation from random mating:

$$\chi^2_3 = 3.584, \quad P = 0.310$$

However, there is a deviation from the expectations of random mating in the matings of intermediates. The numbers of matings,

I × I (141)   I × Not-I (174)   Not-I × Not-I (77)

lead to an estimated preference of intermediates for intermediates of

$$\hat{\beta} = 0.187$$

though $\chi_1^2 = 3.031$

is not quite significant at

$$P = 0.0817$$

We have already estimated that according to Model 4P, the total preference for intermediates is

$$\hat{\beta} = 0.262$$

If the assortative component of the preference can be no more than 20 per cent of the total, then this component of the preference should only be

$$\hat{\beta} = 0.0524$$

Compared with this, the estimated preference is much too high. But it is not in fact significantly different from zero. Low levels of assortative preference can only be detected significantly in very large bodies of data.

When matings are classified as between melanics and non-melanics, we have the matings and numbers

$$M \times M \ (254) \quad M \times P \ (119) \quad P \times P \ (19)$$

where M represents dark or intermediate and P represents pale. In this case I estimate the preference of melanics for melanics to be

$$\hat{\beta} = 0.216$$

This is also a high estimate, but again not significantly different from zero.

Assuming random mating of melanics and pales,

$$\chi_1^2 = 1.070, P = 0.301$$

a typical result if mating were, indeed, truly random. Only very high levels of assortative preference can be detected as statistically significant given the total number of matings shown in table 10.5.

In the period 1946–59, several naturalists observed matings of melanic and non-melanic Arctic Skuas in different Shetland colonies. O'Donald (1960b) analysed these data: the matings, which include matings on Fair Isle 1948–59, show significant assortment which is homogeneous between colonies. Table 10.6 gives all the data now collected together on numbers of matings of Arctic Skuas. The table also shows the model I fitted to the data, the estimates of the assortative preference for melanics and the analysis of χ^2. The model simply assumes a preference β of melanics for melanics. It is just a special case of the general model of assortative mating shown in table 9.10, chapter 9. The results of the analysis are remarkably similar to the

Table 10.6. *Assortative mating of melanic and pale phenotypes of Arctic Skuas*

(i) Model of assortative mating for melanics

Mating	Frequency
Melanic × melanic	$\beta u + u^2(1-\beta)^2/(1-\beta u)$
Melanic × pale	$2u(1-u)(1-\beta)/(1-\beta u)$
Pale × pale	$(1-u)^2/(1-\beta u)$

The M.L. estimates of the parameters β and u of this and other assortative mating models have been given by Davis & O'Donald (1976a).

(ii) Numbers of matings and estimates of preference

Mating	Fair Isle 1973–79	Foula 1975	Shetland 1946–59	Total
Melanic × melanic	254	144	218	616
Melanic × pale	119	86	120	325
Pale × pale	19	26	38	83
Estimate: $\hat{\beta}$	0.2159	0.3898	0.4434	0.3826
Variance: var $(\hat{\beta})$	0.03289	0.01550	0.008964	0.004876

(iii) Analysis of χ^2 when the model is fitted to the data

Component of variation	Value of χ^2	Dfs	Value of P
Preferential mating of melanics with $\hat{\beta} = 0.3826$	19.2720	1	1.13×10^{-5}
Differences in phenotypic frequencies	11.0010	2	0.00408
Residual heterogeneity	1.4882	2	0.475
Total	31.7612	5	6.62×10^{-6}

The values in this table are taken from table C5, appendix C, section C.4.1.

results of fitting Model 3P to the distributions of breeding dates (tables 10.2 and 10.4). The estimates of the mating preference are almost identical. But this must be largely the effect of chance: assuming a roughly Normal distribution of error for the estimate $\hat{\beta}$, its 95% confidence limits are given by

$$0.246 \leq \beta \leq 0.519$$

The assortative mating is estimated to be much smaller on Fair Isle than in the rest of Shetland. Foula and Shetland generally show highly significant assortative mating, giving an estimate of preference

$$\hat{\beta} = 0.422$$

which is twice that for Fair Isle. Surprisingly, perhaps, the difference between Fair Isle and the rest of Shetland is not significant, as the value of

χ^2 for residual heterogeneity shows. But differences between colonies in assortative mating would have to be very large to be detected in samples of the size I have obtained. Differences in the mating systems of the colonies could easily have arisen when the colonies were originally founded, presumably by a few pairs of birds. These differences would tend to persist as the colonies expand because many of the recruits will come from birds hatched and reared in the colony. On Fair Isle only a very few birds are known to have been reared in other colonies. The Fair Isle colony has been in existence for only about four or five generations – too short a time for migration to have obliterated a founder effect.

As I have already emphasized, computer simulations of genetical models show that assortative mating evolves as mating preferences evolve (O'Donald, 1980*a*). But at most, no more than about 20 per cent of the preference is expressed assortatively. Table 10.6 shows that an estimated 38 per cent of melanic females prefer the melanic males. Since about 76 per cent of the birds are melanic, a total of 29 per cent of all females express this preference. But we have also estimated (table 10.2) that 38 per cent of all females mated preferentially, either with or without assortment. The assortative matings thus represent 76 per cent of all preferential matings – a much larger percentage than the maximum value of 20 per cent that is found in the genetical models.

This is not a complete refutation of the theory, however. The data of all Arctic Skua colonies were used to estimate the assortative preference. But the estimate of the total preference came only from the breeding dates of Arctic Skuas on Fair Isle. If we assume that the preferences on Fair Isle really are different from those in the rest of Shetland, we should use the estimate of the assortative preference on Fair Isle to compare with the estimate of the total preference. Given these estimates, we find that the assortative matings represent 46 per cent of all preferential matings on Fair Isle. Since the assortative mating is not in fact statistically significant, this percentage is not significantly different from zero. The large sampling error of the estimate thus serves not to refute the genetical model of the evolution of mating preferences.

10.4 Summary of evidence for female choice

Four lines of evidence support the hypothesis that sexual selection of male Arctic Skuas takes place by female choice, not male competition:

    (i) the goodness of fit of models of mate selection to the distributions of breeding dates;

(ii) the assortative mating between the phenotypes;
(iii) the stability of the polymorphism;
(iv) the effect of male experience on breeding success.

(i) Goodness of fit of models of mate selection

In section 10.2 of this chapter, I described the results of fitting models to the distributions of breeding dates. We saw that the P models in which preferential matings precede random matings gave the best fit. This is consistent with what we should expect if females evolved preferences for melanic males. Models in which random matings came first would be consistent with differences in levels of male courtship. If melanic males were more active in courting the females, they would gain mating advantage with the less responsive and more choosy females who would mate later than the very responsive females choosing any available male they encountered.

(ii) Assortative mating between phenotypes

The assortative mating of melanic phenotypes strongly supports the hypothesis of female choice. Assortative mating could occur by male competition if the males competed only for females like themselves. This seems very implausible. In higher vertebrates, female choice operating through preferences expressed when females possess the same phenotypes as the males is almost certainly the mechanism of assortative mating.

(iii) Stability of polymorphisms

Analysis of genetic models (section 9.2 of chapter 9) shows that female choice necessarily entails frequency-dependent mating. This will always produce a stable polymorphism if more than one male phenotype is the object of the females' preferences. It can also produce stable polymorphisms in equilibrium with natural selection (section 9.2.3). Male competition in which there is no element of female choice – fighting over the females or competing for territories – does not entail frequency-dependent mating. Polymorphisms are maintained only if heterozygous males are the best competitors and thus gain heterozygous advantage.

In the Arctic Skua the heterozygous intermediates are somewhat less successful than the homozygous dark males in finding mates: they take an extra day and a half to find a mate and thus suffer a slight disadvantage by breeding later on average during the season (section 8.5 of chapter 8). There is no possibility in the Arctic Skua that sexual selection might produce a heterozygous advantage.

The polymorphism of the Arctic Skua has shown remarkable stability in Shetland over the period 1934–79 which covers five generations. The phenotypic frequencies show a cline of increasing proportion of pale birds from south to north (section 2.5, chapter 2). Stable clines can certainly be produced by a balance between migration and a gradient of selection from advantage to disadvantage (Fisher, 1950; Haldane, 1948) with no frequency-dependence or heterozygous advantage in fitness values. Arctic Skuas often migrate between colonies. Roughly 50–55 per cent of the adult birds return to breed in their natal colony (section 2.4, chapter 2); 45–50 per cent migrate to other colonies in Shetland. The handful of known birds coming from elsewhere to breed on Fair Isle were born on Foula, Hermaness, Bressay, Mousa and Noss in Shetland, but nowhere further away. There is no evidence of migration between widely separated colonies that vary in phenotypic frequency along the cline. But migration between adjacent colonies in one generation may lead to gene flow further along the cline in subsequent generations. In theory, the cline could be maintained either by a balance between selection and migration or by variation in the balance between sexual and natural selection.

In Model 3P, the sexual selection in favour of melanic males would ultimately produce a homozygous dark population. This sexual selection forms a component of the general reproductive advantage of the dark phenotypes (table 6.9, section 6.3, chapter 6). The pale phenotypes mature about half a year earlier than the melanics and gain an advantage from their greater chance of surviving to breed (section 6.2, chapter 6). Sexual and natural selection are thus in opposition, but not exactly balanced. Darks have the overall advantage: pales should be eliminated unless they are continually replaced by immigration.

(iv) Effect of male experience

Males who are taking a new mate may or may not have had previous breeding experience. Their experience has no effect on their chances of finding a mate: dark males who are forced to find a new mate by the death or divorce of their previous mate are no more successful than dark males coming to breed for the first time (section 8.4.3, chapter 8). If males competed for females by fighting or holding large territories or courting the females, we should surely expect that they would benefit from previous experience of these activities. Older, experienced males should certainly be better fighters. They would already have territories to come back to. They should be better at selling themselves to the females. Previous experience should then show itself in earlier breeding. But it does not.

Experienced and inexperienced dark males breed at almost exactly the same average date, as table 8.4 (section 8.4.3, chapter 8) shows. Experienced intermediates are somewhat earlier on average than inexperienced intermediates and this produces a slightly earlier average breeding date of experienced males in general. These small differences are by no means significant. But statistical tests of significance can only be used to refute a hypothesis, not to confirm it. We can only say that the data of table 8.4 do not support hypotheses based on male competition. Yet if male competition were the cause of sexual selection, male experience should have a strong effect on the chances of mating. Males new to the colony, breeding for the first time, are at quite a severe disadvantage compared to males who have bred before. New males have no territory to come straight back to. They must find and defend a completely new territory. This should delay them. Their territory is likely to be on the edge of the skua colony and away from the scene of most of the action. They will have less time for courtship, fewer females to court. The new territories are indeed often in outlying areas separated from the rest of the colony. In the Fair Isle colony of 1976 (see the map of figure 1.1 in section 1.1, chapter 1), new pairs obtained territories behind Ward Hill, on Ulieshield, Mopul and Vaasetter, all outlying areas. If males do compete for females, experience should count for much of the chances of getting one.

But if male competition and salesmanship count for nothing in the race to find a mate, what does? Female preference is at least consistent with the absence of any apparent effect of previous experience. If females have evolved preferences, they know what they want in a mate. If females prefer dark males, an inexperienced dark male may be as much the object of female choice as an experienced dark male. This further illustrates the distinction I have already made (in section 10.2) between the two mechanisms of female choice: females may choose males because some males are better salesmen than others; they may choose because they have an inherent preference for a particular phenotype regardless of how well the males sell themselves. Perhaps female choice should be used as an inclusive term for both mechanisms, while female preference should only be applied when females have certain inherent preferences to express as in the ordinary use of the term preference.

10.5 The mechanism of sexual selection: conclusion

The evidence for female preference that I have discussed in the previous section is not entirely conclusive, even though it is strongly supported by three independent lines of evidence. The P models of preferential mating are an excellent fit to the data of the breeding dates of the male phenotypes. Assortative mating, which occurs in Arctic Skua populations in Shetland, should evolve as the female preferences evolve. Male competition would not give rise to assortative mating unless males competed more strongly when courting a female with a phenotype like their own. This seems highly implausible, but has not been excluded.

On Fair Isle, sexual selection for the dark phenotype opposes natural selection for the pale. If female preference gives rise to the sexual selection, it could maintain a stable polymorphism. But so could migration along a cline in which the dark phenotypes were advantageous in the south and the pale phenotypes advantageous in the north. Female choice is clearly not entailed by the stability of the polymorphism. But it would be implied if it could be shown that the cline is maintained not by diffusion but by a balance of selective forces.

Female choice can benefit new and experienced males equally. Male competitive ability should improve with experience. Since experienced males are not more successful at finding a mate than new males, this supports the theory of female choice.

These three lines of evidence – the fit of the models, the assortative mating, and the equal success of new and experienced males – all give strong support to the theory of female choice. Explanations in terms of male competition would in each case require additional postulates, some of them quite implausible.

Thus I conclude that female Arctic Skuas prefer to mate with dark and intermediate males who pair and breed at an earlier average date than pale males. They gain a selective advantage over pale males, exactly as Darwin had postulated, because early pairs are more successful than late pairs in the number of fledglings they produce. The advantage they gain in this way applies only to males in new pairs. The selective advantage in the population as a whole depends on the proportion of new pairs and the relative fledging success of new and older pairs. Intermediates are at a slight sexual disadvantage compared to darks and pales at a much greater disadvantage. In section 8.5 of chapter 8, I calculated the coefficients of sexual selection, as follows:

$$s_I = 0.013$$
$$s_P = 0.058$$

These values represent the relative disadvantage of intermediate and pale males compared to dark males.

11

Conclusions

In this book I have described the results of my study of the population ecology and genetics of the Arctic Skuas on Fair Isle. When a population ecologist studies an organism, he asks questions, such as the following, about its population size and numbers:

    (i) Are its populations stable, increasing, or decreasing?
    (ii) Can future population changes be predicted?
    (iii) How are population numbers regulated – for example, is regulation density-dependent?

Seeing a polymorphism, a population geneticist asks such questions as:

    (i) What are the genetics of the polymorphism?
    (ii) Is the polymorphism stable, or is it still evolving?
    (iii) How are the gene frequencies spatially distributed – is their distribution uniform, or are there clines from one area to another?
    (iv) What selective forces are acting on the phenotypes – natural selection, sexual selection, or both – and are they sufficient to 'protect' the polymorphism against extinction of alleles?

I hoped I might answer such questions when I set out to study the Arctic Skua and its striking polymorphism in plumage. In previous chapters – chapter 2 on population ecology, chapter 5 on genetics and chapters 6, 8 and 10 on natural and sexual selection, I have given some answers. Some answers are tentative: some – on the population changes and the natural and sexual selection of the population – are bold and decisive. Now, in this last chapter, I can omit the details of analysis, summarize the results, and draw general conclusions about the evolutionary forces that maintain the polymorphism.

We saw, in chapter 2, section 2.2, that Arctic Skua populations have either increased in recent years, as on Fair Isle and Foula, or remained roughly stable, as on Noss and Unst. As in all studies of population ecology, the validity of these conclusions is entirely dependent on the census and survey methods used to estimate the population numbers. On Fair Isle, each nest has been marked and mapped in the course of detailed surveys of the Arctic Skua colony repeated at intervals throughout the breeding season. In recent years, the smaller colony on Noss has also been surveyed throughout the breeding season and each nest marked. These methods give the exact numbers of pairs breeding each year: by repeated surveying, all pairs are eventually found. In 1975, the Arctic Skuas of Foula were also surveyed throughout the breeding season. Each nest was marked when it was found. This produced a much larger estimate of the number of pairs than previous estimates obtained from surveys or counts made at particular times in the breeding season. Unless an Arctic Skua colony is surveyed repeatedly in the period when the birds have eggs or chicks, many pairs are certain to be missed and estimates of population size are unlikely to be comparable between one year and another. Nevertheless, estimates of numbers of Foula do show a long-term upward trend in spite of considerable variation between different years

On Fair Isle, in the years 1948–62, pairs of Arctic Skuas increased at an exponential rate of about 11 per cent from one year to the next. In 1973–75, they also increased rapidly. But then, from 1976 onwards, increasing numbers were recorded as having been shot. The annual mortality increased from 11 per cent in 1973–75 to 31 per cent in 1978. At this rate, the Arctic Skua colony on Fair Isle will probably have been exterminated in about 30 years' time. In section 6.1.1 of chapter 6 (on demography and selection), I calculated the demographic effect of the increased mortality by shooting. It will produce an intrinsic rate of decrease

$$r = -0.170$$

and hence an annual exponential decrease of

$$100(1 - e^r) = 16\%$$

At this rate, the Arctic Skua colony will be reduced from 114 pairs in 1979 to 21 pairs in 1989 and 4 pairs in 1999.

The calculation of the rate of decrease depends on the assumption that only about 45–55 per cent of surviving Arctic Skua chicks return to breed in their natal colony. Immigrants from neighbouring colonies make up the remainder of the birds that come to breed on Fair Isle (section 2.4, chapter 2). Immigration roughly balances emigration and will have no net effect on the population demography. Although subject to sampling error (the mortality of 31 per cent has a standard error of 3.3 per cent), the effect of shooting is always to produce a large negative value of r: the present rate of mortality will exterminate the Arctic Skuas of Fair Isle. But the mortality may to some extent be density-dependent. The Arctic Skuas appear to have annoyed some people more as the skuas increased and expanded over a greater proportion of Fair Isle. If the increased shooting is partly a response to the increased numbers, a smaller proportion may be shot as the colony contracts in size. There is no evidence that the Arctic Skuas were shot when the colony consisted of just a few pairs. The final outcome will depend on whether anybody is really determined to exterminate the skuas and whether action to prevent this is taken.

In other parts of Shetland, where human action has produced no deliberate increase in mortality, Arctic Skua populations are stable or increasing, as on Foula, Noss and Unst. But these three populations suffer from competition with Great Skuas or Bonxies. Since 1968, Bonxies have increased at an annual exponential rate of about 14 per cent on Foula where they take about 30 per cent of the Arctic Skuas' fledglings. This extra mortality should be large enough to start producing a slow decline of the Arctic Skuas on Foula (see section 2.3, chapter 2). At present, this is only a predicted decline: it has not actually occurred. But as Bonxies continue to increase, the Arctic Skuas cannot also continue to increase for long. On the nature reserve of Hermaness on Unst, Arctic Skuas were very common – about 200 pairs nested in 1922 – but rapidly declined to about 70–80 pairs after 1950. They now appear to have been forced off Hermaness on to other nesting grounds on Unst. This has occurred partly because Bonxies always win in competition for territories. They come back to the breeding grounds in the spring before the Arctic Skuas. A pair of Arctic Skuas cannot dislodge a Bonxie who has already settled in the Arctic Skuas' former territory. As the Bonxies expand their colony, they drive away the Arctic Skuas.

No evidence suggests that the Bonxies' predation and competition might be density-dependent. On the contrary, the larger an Arctic Skua colony, the more it is protected against intruding Bonxies by the fighter squadrons of Arctic Skuas. Isolated pairs of Arctic Skuas are highly vulnerable to nearby Bonxies. If anything, the Bonxies' predation would be negatively density-dependent, increasing at lower densities and accelerating the rate of population decline. In populations where no Bonxies or humans compete with the Arctic Skuas, I can only presume that food or space will ultimately limit the population growth. Since Arctic Skuas need spend so little of their time hunting for food to gain the energy they need (section 3.3, chapter 3), populations must still be well below the point at which food would become limiting.

To a population geneticist studying a polymorphism, the last of my questions is the most important: is the polymorphism 'protected'? If so the alleles that determine the different phenotypes are unlikely to be lost from the population. It is analogous to the ecologist's question: what regulates the population size? If regulation is density-dependent, a population is unlikely to become extinct. Suppose that if an allele were near the point of extinction, natural or sexual selection would always act to raise its frequency. Extinction would then be an unstable point. The allele would be protected: it could only be lost by chance from a small population. Every allele, and hence the polymorphism, is protected when all extinction points are unstable. The conditions for protection determine the qualitative outcome of the evolutionary process: will the polymorphism be maintained in the population or not? The conditions for protection depend both on the mating system of preferential and random mating and on the coefficients of natural selection. Unfortunately, although some of the components of selection can often be measured in populations, estimates of the total effects of selection are very difficult to obtain. In spite of many years' work by many investigators studying selection of the phenotypes of the snail *Cepaea nemoralis*, the maintenance of the polymorphism is still not understood; though a great body of knowledge has been produced on various aspects of selection in different populations and environments.

When all individuals in a population can be marked, and their survival and reproduction followed throughout their lives, intrinsic rates of increase can be calculated for different phenotypes and hence the effects of selection upon them. As organisms for such demographic studies, birds are particularly convenient: they are often easier to catch and mark than other organisms; all their nests can be found and their reproductive rates observed. A ground-nesting bird like the Arctic Skua, which always returns to the same nesting ground, must be close to the ideal for demographic studies. From data of the survival and reproduction of the pale, intermediate and dark phenotypes, I estimated their intrinsic rates of increase and hence coefficients of selection (section 6.3, chapter 6). Sexual selection by female choice gives rise to one of the components of the selection of the males (section 8.5, chapter 8). But the selection thus estimated will not protect the polymorphism. The intermediate heterozygotes suffer an overall disadvantage. Both fixation states are stable: one allele must be eliminated while the other is fixed. The selection also differs between the sexes: dark males have the advantage over pale males; pale females have the advantage over dark females. A sex difference can produce a stable polymorphism within a wide range of selection parameters. But not in the Arctic Skua. The overall heterozygous disadvantage ensures that the fixation states are stable. An unstable polymorphic equilibrium exists at the point when the allele for melanism has the frequency

$$p_* = 0.2932$$

Above this point, the melanic allele will continue to increase in frequency until it has reached fixation and the pale allele has been eliminated. On Fair Isle, about 80 per cent of Arctic Skuas are either dark or intermediate. The frequency of the dark allele is roughly

$$p = 0.55$$

This is well above the unstable equilibrium point. Dark birds should eventually become fixed and pale birds eliminated. Yet, in the rest of Shetland, where the frequency of pale birds is slightly higher than on Fair Isle, their frequency has remained almost constant at 26 per cent for at least five generations. Over this period of time, the selection parameters estimated in table 6.10 (section 6.3, chapter 6) should have reduced the frequency of the pale birds to 24 per cent. This change is about equal to the standard error in the estimate of frequency. Not enough time has passed for the predicted change in frequency to become detectable statistically.

In theory, as shown in section 9.2.3 (chapter 9), the natural selection against dark and intermediate birds produced by their later age at maturity can be balanced by the sexual selection in their favour. The balance is a consequence of the frequency-dependence of sexual selection by female choice (explained in section 9.3, chapter 9). As the preferred males increase in frequency, their sexual advantage declines until it may exactly balance their natural selective disadvantage. But a computer simulation shows that a balance would not be reached in the Fair Isle population of the Arctic Skua. The melanic phenotypes always gain the overall advantage. Using Model 3 and its estimates of the female preferences (sections 10.1 and 10.2, chapter 10), I computed the sexual advantage that the melanic males would gain as they increased in frequency. From their actual frequency on Fair Isle to the point of complete fixation, they show only a slight frequency-dependent decline in advantage – about two per cent. This is much too small to produce a point of balance with natural selection: the pale birds should still be eliminated from the population. As we have seen, however, the apparent stability of the polymorphism in Shetland is no refutation of this conclusion: the small predicted increase in the melanics frequency is roughly the same as the standard deviation of their observed frequency. About ten generations would probably be necessary to show that an evolutionary change had occurred. The question about the stability of the polymorphism cannot be answered by direct and relatively short-term observations.

The answer to the question about spatial distribution is simple: pale birds form a cline of increasing frequency from about 20 per cent in the south to 100 per cent in the north. The cline is not uniform in relation to latitude (see figure 2.6, section 2.5, chapter 2): in Finland only four per cent are non-melanic pale birds. The cline is certainly not stable everywhere: in Iceland melanics are increasing in frequency; the cline is moving further north. Three alternative models have been proposed to explain a cline:

> (i) The cline represents the wave of advance as advantageous genes spread through a population (Fisher, 1937).
>
> (ii) Selection against disadvantageous genes in one area is balanced by their immigration from other areas where they are advantageous, thus producing a stable diffusion cline (Haldane, 1948; Fisher, 1950; Karlin & Richter-Dyn, 1976).
>
> (iii) The cline is produced simply by a balance of selective forces that vary along the cline. The equilibria vary as the selective balance varies.

I have already rejected the possibility that the cline of the Arctic Skua represents a wave of advance of the gene for melanism. The cline is circumpolar (figure 2.6, section 2.5, chapter 2). To be a wave of advance, the same melanic mutation must have occurred more or less simultaneously in all southern populations and thence spread northwards. This is obviously absurd. In previous discussions in this book (section 2.5, chapter 2, and section 10.4, chapter 10), I left undecided which of the other two models, (ii) or (iii), might apply to the cline. We can now come to a conclusion. The components of selection acting within the population on Fair Isle do not give rise to stable equilibrium, nor would the frequency-dependence of sexual selection be sufficient to do so. The selective balance theory of a cline (iii) can therefore be rejected. The cline can only be a diffusion cline. If so, migration must carry the genes along the cline from areas where they are advantageous into areas where they are disadvantageous. The gene for pale must diffuse into the populations in Shetland where melanism is more common and at selective advantage. Evidently, the melanics' advantage declines further north turning to disadvantage where the pales become the more abundant phenotype. In theory (Karlin & Richter-Dyn, 1976), the 50 per cent point of frequency (both melanics and pales at equal frequencies) should occur in a neutral zone where neither phenotype has the advantage. Figure 2.7 (section 2.5, chapter 2) shows that the neutral zone occurs at about 67°N. Melanics should be advantageous south of this point and disadvantageous north of it. This is obviously a testable prediction.

Is the migration likely to be sufficient to maintain the diffusion cline? Does the gene for pale diffuse from the north into the Shetland populations? I have no evidence that immigrants come to Fair Isle from further afield than other Shetland colonies. But a lot do come – roughly 45–55 per cent of new breeding birds must be immigrants from elsewhere (section 2.4, chapter 2). In Karlin and Richter-Dyn's model, migration takes place only between adjacent populations. And this is sufficient for genes to flow along the cline. Even though, in one generation, migration takes place only between adjacent colonies of Arctic Skuas, in subsequent generations, genes diffuse between more widely separated colonies; eventually they flow the whole length of the cline. For example, if colonies in the Northern Isles of Shetland receive immigrants from colonies in Faeroe and Norway further north, the genes from the Faeroese or Norwegian colonies can then be passed in the next generation to Shetland colonies further south. Migration rates of about 50 per cent between adjacent colonies will produce high rates of gene flow. This will balance the selective advantage of melanism in Shetland.

The cline is not completely stable: as we have seen, melanics are increasing in Iceland. This is near the 50 per cent point where the most rapid change in clinal frequency occurs. Any movement in the cline will be seen most easily in this region. If the fitness of melanics has increased along the cline, the neutral point will have moved north. The cline will move to follow it. Eventually, the clinal frequencies – the S-shaped curve shown in figure 2.7 – will stabilize further north. But selective forces change with the environment. A cline must always be in some slight movement. Even when the movement is sufficient to be detected, it will have very small effects at the ends of the cline where the S-shaped curve flattens out. In these regions – in Shetland and Orkney in the south and in the far north – very little change in frequency should occur. Local movement in the cline does not refute the general theory of our conclusion. The polymorphism of the Arctic Skua is maintained by a balance of selection and diffusion.

Rates of increase of bird populations

A.1 Estimation of exponential rates of increase

In a population in which birth and death rates depend on age, individuals' ages will eventually reach a stable distribution, which remains the same from one interval of time to the next. When this has happened, the birth and death rates per head of population remain constant: although each age class has its own birth and death rates, the proportions of individuals in each age class are constant. Therefore, in one interval of time, the change in population size (or the size of any age class) is given by

$$N_t = \lambda N_{t-1} = e^r N_{t-1}$$

or for x units of time

$$N_t = e^{rx} N_{t-x}$$

Birds have a discrete, annual breeding cycle. Suppose they survive to age x with probability l_x; having survived, they produce b_x fledglings. On average, the contribution of males and females is equal; hence b_x is equal to half the number of fledglings produced by a mated pair. Individuals who were fledged x years prior to year t will produce $l_x b_x$ fledglings at t, since a proportion l_x of them survived and then produced b_x fledglings. Suppose that in year $t-x$ a total number of

$$N_{t-x} = e^{-rx} N_t$$

fledglings were produced. At time t, x years later, when each of these fledglings produce an average of $l_x b_x$ fledglings themselves, they produce a total of

$$N_t e^{-rx} l_x b_x$$

fledglings. Now these are produced by birds who were fledglings in year $t-x$. The total number of fledglings produced by birds of all ages, $x = 1$, $2, \ldots$, is therefore

$$N_t = \sum_x N_t e^{-rx} l_x b_x$$

so that we obtain Lotka's equation

$$\sum_x e^{-rx} l_x b_x = 1$$

Given values of l_x and b_x, this equation can be used to estimate the instantaneous rate of population increase or decrease r, and hence λ.

Rates of population growth can be measured by r, or λ, or by the increase or decrease over the period of one generation, R_0, where

$$R_0 = N_{t+T}/N_t = e^{rT}$$

T is the mean duration of a generation. Since R_0 is the average number of offspring that an individual contributes to the next generation, it can be calculated by the equation

$$R_0 = \sum_x l_x b_x$$

Thus, knowing r, we can find the generation time

$$T = [\ln(R_0)]/r$$

The value of r is usually estimated by a power series, which is derived as follows. If we put

$$y(r) = \sum_x e^{-rx} l_x b_x$$

then

$$\frac{dy}{dr} = -\left[\frac{\sum_x xe^{-rx} l_x b_x}{\sum_x e^{-rx} l_x b_x} \right] \sum_x e^{-rx} l_x b_x$$

$$= -A(r)y(r)$$

Therefore

$$\frac{dy}{y} = -A\,dr$$

$$y = R_0 e^{-\int A dr}$$

since $y_0 = \sum_x l_x b_x = R_0$. If we put

$$R_1 = \sum_x x l_x b_x$$

$$R_2 = \sum_x x^2 l_x b_x$$

and in general

$$R_j = \sum_x x^j l_x b_x$$

then since

$$A = \left[\sum_x xe^{-rx}l_x b_x\right]\Big/\left[\sum_x e^{-rx}l_x b_x\right]$$

therefore

$$A = \frac{R_1 - rR_2 + r^2 R_3/2 - r^3 R_4/3! + \cdots}{R_0 - rR_1 + r^2 R_2/2 - r^3 R_3/3! + \cdots}$$

$$= \left[\frac{R_1}{R_0} - r\frac{R_2}{R_0} + \frac{r^2}{2}\frac{R_3}{R_0} - \cdots\right]\left[1 - r\frac{R_1}{R_0} + \frac{r^2}{2}\frac{R_2}{R_0} - \cdots\right]^{-1}$$

After expanding the second term in a negative binomial series, multiplying the expressions together, and collecting the terms in $r^0, r^1, r^2, \ldots$, we obtain the series

$$A = \alpha + r\beta + r^2\gamma + r^3\delta + \cdots$$

where

$$\alpha = \frac{R_1}{R_0}$$

$$\beta = \frac{R_1^2}{R_0^2} - \frac{R_2}{R_0}$$

$$\gamma = \frac{R_1^3}{R_0^3} - \frac{3R_1 R_2}{2R_0^2} + \frac{R_3}{2R_0}$$

$$\delta = \frac{R_1^4}{R_0^4} - \frac{2R_1^2 R_2}{R_0^3} + \frac{2R_1 R_3}{3R_0^2} + \frac{R_2^2}{2R_0^2} - \frac{R_4}{6R_0}$$

Given the values of l_x and b_x, the values of $R_0, R_1, \ldots, R_j$ can be calculated and hence the coefficients α, β, γ and δ. The equation for r is then

$$R_0 e^{-\int A dr} = 1$$

or

$$\ln(R_0) = \alpha r + \frac{1}{2}\beta r^2 + \frac{1}{3}\gamma r^3 + \frac{1}{4}\delta r^4 + \cdots$$

If r is a small quantity, terms in r^2 and higher powers of r can be neglected. Then

$$r = [\ln(R_0)]/[R_1/R_0]$$

and

$$T = R_1/R_0$$

This value of the generation time, T, is simply the mean age of bearing offspring, since it is the same as

$$T = \sum_x x \left[\frac{l_x b_x}{\sum_x l_x b_x} \right]$$

and $l_x b_x / \sum_x l_x b_x$ is the proportion of offspring produced at age x. More accurately, we should consider additional terms of the power series in r. To a second degree of approximation, we have the equation

$$\ln(R_0) = \alpha r + \tfrac{1}{2} \beta r^2$$

which is easily solved to give the estimate

$$r = \frac{-1 + [1 + 2(1 - R_0 R_2 / R_1^2) \ln(R_0)]^{\frac{1}{2}}}{R_1 / R_0 - R_2 / R_1}$$

and hence T by the equation

$$T = [\ln(R_0)]/r$$

Many programmable calculators are now provided with programmes to find the roots of polynomials. More accurate estimates for r can easily be obtained by including more terms of the power series. It is not often necessary, however, to go beyond the second term of the series to get an accurate estimate. Dublin & Lotka (1925) originally devised this general method of calculating the rate of population increase.

Values of r can be used to calculate selective values or fitnesses of different genotypes or phenotypes. We need to know l_x and b_x for each genotype or phenotype. Then we can find the values of r for each genotype or phenotype. A convenient measure of selection is the relative fitness

$$w = e^{r_1 T} / e^{r_2 T}$$

where r_1 and r_2 are the rates of increase of two genotypes or phenotypes and T is the mean generation time. The relative fitness is thus the change in relative proportion over a period of one generation. Selection can also be measured by the selective coefficient

$$s = 1 - w$$

where $w < 1$ and is measured relative to the fitness of the most fit genotype or phenotype. Therefore s measures the relative disadvantage of some genotype or phenotype in comparison with the most fit genotype or phenotype.

A.2 Estimation of survival and reproductive rates

The first step in the calculation of values of r is of course to find the values of l_x and b_x for individuals in the population. These values can be observed directly by following a cohort of individuals born in a particular year. Unfortunately, in the Arctic Skua colony on Fair Isle, the numbers of fledglings produced in a year are too small to give worthwhile estimates of l_x. The distribution of ages of breeding birds does not give the values of l_x, since the old birds were fledged when the colony was small and the young birds after it had greatly expanded. The older birds thus come from a smaller proportion of fledglings. In a previous demographic analysis of the Arctic Skua colony on Fair Isle (O'Donald & Davis, 1976), I had attempted to allow for this bias by dividing the number of birds in each age class by the relative size of the colony from which that age class had been drawn. This method assumes that the number of fledglings produced is proportional to the number of birds in the colony. But fledging success varies between years. Some years are more successful than others. The method can also produce the absurdity that, since the birds are not a cohort, estimated values of l_x can actually increase, as when, for example, only one bird in the colony is 13 years old and two birds are 14 years old. This result occurs by chance among the small numbers of very old birds. But it has little effect on the estimate of r, since the values of R_0, R_1 and R_2 are largely determined by the much greater numbers of younger birds.

An alternative method of calculating the values of l_x is to use the annual mortality of the birds. After the first year of life, mortality is usually constant from one year to the next (Bulmer & Perrins, 1973). Suppose a proportion c die in the first year of life and d die each year thereafter. Then the values of l_x form a geometric series

$$l_x = (1-c)(1-d)^{x-1}$$

The values of b_x are estimated by the average number of fledglings produced by birds of different ages. This gives the table of values

Age	Surviving proportion	Reproductive rate
0	$l_0 = 1$	0
1	$l_1 = (1-c)$	b_1
2	$l_2 = (1-c)(1-d)$	b_2
3	$l_3 = (1-c)(1-d)^2$	b_3
4	$l_4 = (1-c)(1-d)^3$	b_4
.	.	.
.	.	.
.	.	.
x	$l_x = (1-c)(1-d)^{x-1}$	b_x

Arctic Skuas breed for the first time aged 3, 4, 5, 6 or 7 years. Most breed first at 4 years. For a bird breeding for the first time at 6 years old,

$$b_1 = b_2 = b_3 = b_4 = b_5 = 0$$

The first year is very significantly less successful than subsequent years. Pale phenotypes in their first year of breeding produce on average only half the fledglings they will produce in later years. After the first year a roughly constant number of fledglings is produced each year.

A four-year-old Arctic Skua breeding for the first time produces b_4 fledglings on average. Suppose that from the fifth year onwards it produces

$$b_5 = b_6 = b_7 = \ldots = b \text{ fledglings.}$$

Then the values of R_0, R_1 and R_2 can be calculated as follows.

$$\begin{aligned}
R_0 &= b_4(1-c)\,(1-d)^3 + b(1-c)\,(1-d)^4[1+(1-d)+ \\
&\quad (1-d)^2+(1-d)^3+\ldots] \\
&= (1-c)\,(1-d)^3[b_4+b(1-d)/d] \\
R_1 &= 4b_4(1-c)\,(1-d)^3 + b(1-c)\,(1-d)^4[5+6(1-d)+ \\
&\quad 7(1-d)^2+8(1-d)^3+\ldots] \\
&= (1-c)\,(1-d)^3[4b_4+b(1-d)\,(4/d+1/d^2)] \\
R_2 &= 16b_4(1-c)\,(1-d)^3 + b(1-c)\,(1-d)^4[25+36(1-d)+ \\
&\quad 49(1-d)^2+64(1-d)^3+\ldots] \\
&= (1-c)\,(1-d)^3[16b_4+b(1-d)\,(16/d+7/d^2+2/d^3)]
\end{aligned}$$

Suppose in general that a bird has a constant fledging rate, b, from its kth year. From then onwards it contributes R_0', R_1' and R_2' to the total values of R_0, R_1 and R_2. It can be shown that

$$\begin{aligned}
R_0' &= b(1-c)\,(1-d)^{k-1}/d \\
R_1' &= b(1-c)\,(1-d)^{k-1}[(k-1)/d+1/d^2] \\
R_2' &= b(1-c)\,(1-d)^{k-1}[(k-1)^2/d+(2k-3)/d^2+2/d^3]
\end{aligned}$$

If a bird breeds for the first time in its kth year and has a constant reproductive rate throughout life, then of course we have

$$\begin{aligned}
R_0 &= R_0' \\
R_1 &= R_1' \\
R_1 &= R_2'
\end{aligned}$$

To a first approximation

$$T = R_1/R_0 = k-1+1/d$$

This is what we should expect: $k-1$ is the number of years up to breeding age and $1/d$ is the expectation of the breeding life.

A.3 Rates of increase of Arctic Skuas on Fair Isle and Foula
On Foula, Arctic Skuas suffer in competition with Bonxies. The

Bonxies occupy their territories and prey on their chicks. As shown in chapter 2, section 2.3, about 30 per cent of the chicks are predated each year. Calculation of r shows that this competition should produce a decline in the population on Foula.

In the population on Fair Isle, the average survival and reproductive rates are roughly

$d = 0.2$

$b = 0.6$ (1.2 fledglings per pair)

For many seabirds, the mortality in the first year is about

$c = 0.3$

We have estimated (chapter 2, section 2.4) that at least 45 per cent of chicks return to breed in their natal, Fair Isle colony. In the year 1973, about 35 chicks were fledged. If all had survived, 61 would have returned to Fair Isle to breed, but only 21, about 34 per cent, did so. At four years old, when most birds breed for the first time, the probability of surviving is

$(1-c)(1-d)^3 = 0.34$

If $d = 0.2$, then

$c = 0.32$

Thus we may assume, roughly, that

$c = 0.3$

$d = 0.2$

$b = 0.6$

and that birds breed at four years old.

Then

$$R_0 = b(1-c)(1-d)^3/d = 1.08$$
$$R_1 = b(1-c)(1-d)^3[3/d+1/d^2] = 8.60$$
$$R_2 = b(1-c)(1-d)^3[9/d+5/d^2+2/d^3] = 90.3$$

To the first approximation

$$r = \ln(R_0)/(R_1/R_0)$$
$$= 0.00906$$

To the second approximation

$$r = \frac{-1+[1+2(1-R_2R_0/R_1^2)\ln(R_0)]^{\frac{1}{2}}}{R_1/R_0-R_2/R_1}$$

$$= 0.00917$$

On Fair Isle, Arctic Skuas should be increasing each year at an exponential rate of $100(e^r-1) = 0.9$ per cent. This low rate of increase is reasonable since the population numbers have been more or less stable after 1976. At low rates of increase, the first approximation is sufficiently accurate for the calculation of r.

On Foula, the chicks suffer an additional 30 per cent mortality from predation. Their probability of surviving predation is roughly 0.7. Their probability of then surviving for the whole of their first year of life is also roughly 0.7. Therefore their overall mortality in their first year is

$$c = 1-(0.7)^2$$
$$= 0.51$$

Using this value with

$$d = 0.2$$
$$b = 0.6$$

as for Fair Isle, we obtain

$$R_0 = 0.753$$
$$R_1 = 6.02$$
$$R_2 = 63.2$$

and hence to the first approximation

$$r = -0.0355$$

or to the second approximation

$$r = -0.0341$$

The population should be declining at 3.3 per cent each year. Yet the population appeared to have increased rapidly from 1973 to 1975, from 130 pairs counted in 1973 to the nests of 278 pairs mapped in 1975. I have argued in chapter 2, section 2.2, that much of this increase was only apparent, for the 278 pairs counted by mapping the nests were found by an exhaustive survey carried on throughout the breeding season. Earlier surveys were carried out over short periods and were certainly not exhaustive. Nevertheless, some increase in numbers may have occurred. But the effects of increased predation would not have begun to affect recruitment into the adult population on Foula until about 1975. Only in later years should the population show evidence of decline on the basis of these calculations.

The estimates of intrinsic rates of increase have been derived by making two assumptions: the birds come to maturity and breed for the first time at four years old; they produce an average of 0.6 chicks in each year of their breeding life. Both these assumptions are false. The birds vary in their age at maturity. The average age at maturity is 4.4 years. An individual parent bird produces an average of 0.4113 chicks in its first year of breeding

Table A1. *Expressions for the calculation of the values of R_0, R_1 and R_2*

Age at maturity (years)	R_0	R_1	R_2
3	$A_3b^* + A_4b/d$	$3A_3b^* + A_4b(3/d + 1/d^2)$	$9A_3b^* + A_4b(9/d + 5/d^2 + 2/d^3)$
4	$A_4b^* + A_5b/d$	$4A_4b^* + A_5b(4/d + 1/d^2)$	$16A_4b^* + A_5b(16/d + 7/d^2 + 2/d^3)$
5	$A_5b^* + A_6b/d$	$5A_5b^* + A_6b(5/d + 1/d^2)$	$25A_5b^* + A_6b(25/d + 9/d^2 + 2/d^3)$
6	$A_6b^* + A_7b/d$	$6A_6b^* + A_7b(6/d + 1/d^2)$	$36A_6b^* + A_7b(36/d + 11/d^2 + 2/d^3)$
7	$A_7b^* + A_8b/d$	$7A_7b^* + A_8b(7/d + 1/d^2)$	$49A_7b^* + A_8b(49/d + 13/d^2 + 2/d^3)$

In this table b^* is the reproductive rate of individuals breeding for the first time; b is the reproductive rate of individuals in the second and later years of their breeding lives. $A_j = (1-c)(1-d)^{j-1}$ for values $j = 3, \ldots, 8$.

(0.8226 chicks as a pair). In subsequent years, each parent produces 0.6696 chicks on average (table A2). Given the following distribution of ages (years) at maturity (section 6.2)

Age at maturity;	3	4	5	6	7
No. of birds	13	47	30	10	1

the average probability of survival to maturity now becomes

$$(1-c)(1-d)^2 [13 + 47(1-d) + 30(1-d)^2 + 10(1-d)^3 + (1-d)^4]$$

This must be equal to 0.346, the actual proportion of birds that returned to breed. Therefore, if $d = 0.2$, we obtain the estimate

$$c = 0.275$$

If we use the more accurate estimate,

$$d = 0.1994$$

then

$$c = 0.2768$$

Table A1 gives the general expressions for the values of R_0, R_1 and R_2 for birds reaching maturity at ages 3, 4, 5, 6, and 7 years. Having calculated these values for each age, the average values of R_0, R_1 and R_2 are then calculated from the distribution of ages at maturity. For the population of Arctic Skuas on Fair Isle, the values are

$$R_0 = 1.0724$$
$$R_1 = 9.2055$$
$$R_2 = 101.59$$

Table A2. *Total numbers of chicks fledged to individual birds breeding in successive years in the colony*

Phase	Exp.	Fledging success of males			Fledging success of females		
		No.	Sum	Mean	No.	Sum	Mean
Pale	1	22	15	0.6818	29	17	0.5862
	2	18	20	1.1111	22	27	1.2273
	3	11	15	1.3636	19	22	1.1579
	4	10	11	1.1000	16	24	1.5000
	5	6	6	1.0000	14	18	1.2857
	6	2	3	1.5000	10	13	1.3000
	7	2	3	1.5000	8	10	1.2500
	8	2	4	2.0000	5	9	1.8000
	9	2	2	1.0000	3	2	0.6667
	10	2	3	1.5000	1	1	1.0000
	11	1	2	2.0000	1	2	2.0000
	12	1	0	0.0000	1	1	1.0000
	13	1	2	2.0000	1	2	2.0000
	14	—	—	—	1	1	1.0000
I & DI	1	71	58	0.8169	70	63	0.9000
	2	48	63	1.3125	54	63	1.1667
	3	40	37	0.9250	43	54	1.2558
	4	30	39	1.3000	33	46	1.3939
	5	24	30	1.2500	25	30	1.2000
	6	19	28	1.4737	18	32	1.7778
	7	12	20	1.6667	13	22	1.6923
	8	8	11	1.3750	8	14	1.7500
	9	5	10	2.0000	5	10	2.0000
	10	2	2	1.0000	4	5	1.2500
	11	2	4	2.0000	3	4	1.3333
	12	1	2	2.0000	2	2	1.0000
	13	—	—	—	2	4	2.0000
	14	—	—	—	1	2	2.0000
Dark	1	32	28	0.8750	24	23	0.9583
	2	24	31	1.2917	19	23	1.2105
	3	22	34	1.5454	15	19	1.2667
	4	20	31	1.5500	10	13	1.3000
	5	15	25	1.6667	9	13	1.4444
	6	11	14	1.2727	5	6	1.2000
	7	8	13	1.6250	3	6	2.0000
	8	5	9	1.8000	2	3	1.5000
	9	3	4	1.3333	1	2	2.0000

In the first column of the table referring to phase, the symbol I+DI refers to intermediate and dark-intermediate birds. The column Exp. refers to breeding experience in years from the first year the birds bred in the colony. Sum and Mean refer to the total and mean numbers of chicks produced.

and hence the intrinsic rate of increase

$$r = 0.008225$$

This is less than the estimate previously obtained. For the Arctic Skuas on Foula

$$R_0 = 0.7507$$
$$R_1 = 6.4438$$
$$R_2 = 71.112$$

and hence

$$r = -0.03214$$

This is slightly greater than the previous estimate for Foula, indicating a reduced rate of population decline.

Table A3. *Breeding dates and chicks fledged in period 1948–62*

(i) Females

Type	Year	Breeding date			No. chicks fledged		
		No.	Mean	Variance	No.	Mean	Variance
P	1	34	31.471	61.105	52	0.692	0.609
P	2	28	24.536	26.480	33	1.303	0.655
P	3	13	22.462	19.936	18	1.278	0.448
P	4	13	19.615	19.423	14	1.643	0.401
P	5	5	20.000	22.500	7	1.429	0.952
P	6	6	18.833	31.767	6	1.500	0.300
P	7	4	20.500	53.667	5	1.400	0.800
P	8	1	16.000	0	2	2.000	0
P	9	0	0	0	2	0.500	0.500
P	10	1	16.000	0	1	1.000	0
NOTP	1	139	29.345	64.605	197	0.970	0.683
NOTP	2	82	24.463	32.375	105	1.267	0.601
NOTP	3	56	22.929	28.613	68	1.338	0.406
NOTP	4	36	21.028	19.685	43	1.279	0.539
NOTP	5	23	19.870	22.664	27	1.481	0.413
NOTP	6	17	18.882	16.360	18	1.611	0.487
NOTP	7	8	19.250	6.214	10	1.900	0.100
NOTP	8	6	18.000	16.000	8	1.500	0.571
NOTP	9	4	15.500	15.000	4	2.000	0
NOTP	10	1	18.000	0	2	1.500	0.500
I	1	74	29.689	65.779	104	0.990	0.709
I	2	39	24.359	26.236	51	1.294	0.652
I	3	29	23.966	29.820	34	1.382	0.365
I	4	16	20.500	17.333	19	1.316	0.561
I	5	7	18.714	20.571	10	1.400	0.489
I	6	5	18.800	10.700	5	2.000	0

Table A3 (*cont.*)

(i) Females

Type	Year	Breeding date			No. chicks fledged		
		No.	Mean	Variance	No.	Mean	Variance
I	7	3	21.333	4.333	3	1.667	0.333
I	8	2	16.000	2.000	2	1.500	0.500
I	9	2	15.500	4.500	2	2.000	0
I	10	0	0	0	1	1.000	0
I & DI	1	101	29.099	66.510	141	0.965	0.677
I & DI	2	54	23.704	28.401	71	1.338	0.656
I & DI	3	39	22.923	28.231	46	1.413	0.337
I & DI	4	24	20.292	14.303	30	1.333	0.575
I & DI	5	12	18.417	19.174	16	1.438	0.396
I & DI	6	10	17.900	8.544	10	1.900	0.100
I & DI	7	5	19.800	7.700	6	1.833	0.167
I & DI	8	5	17.200	15.200	5	1.800	0.200
I & DI	9	4	15.500	15.000	4	2.000	0
I & DI	10	1	18.000	0	2	1.500	0.500
DI	1	20	28.250	75.461	29	0.828	0.576
DI	2	9	22.222	37.444	13	1.538	0.603
DI	3	9	20.222	13.194	9	1.444	0.278
DI	4	5	20.800	6.700	8	1.125	0.696
DI	5	3	16.667	20.333	4	1.500	0.333
DI	6	3	16.000	3.000	3	2.000	0
DI	7	0	0	0	1	2.000	0
DI	8	1	15.000	0	1	2.000	0
DI	9	1	11.000	0	1	2.000	0
DI	10	0	0	0	0	0	0
DI & D	1	44	30.205	67.701	67	0.925	0.646
DI & D	2	26	24.962	40.438	35	1.257	0.550
DI & D	3	20	21.000	12.632	22	1.318	0.418
DI & D	4	15	22.067	24.638	18	1.167	0.500
DI & D	5	13	20.846	24.808	14	1.571	0.418
DI & D	6	9	18.778	25.194	10	1.400	0.711
DI & D	7	2	18.000	8.000	4	2.000	0
DI & D	8	2	18.500	24.500	4	1.250	0.917
DI & D	9	1	11.000	0	1	2.000	0
DI & D	10	0	0	0	0	0	0
D	1	22	31.227	58.470	35	1.029	0.734
D	2	17	26.412	38.007	20	1.050	0.471
D	3	11	21.636	12.455	13	1.231	0.526
D	4	10	22.700	34.011	10	1.200	0.400
D	5	10	22.100	20.989	10	1.600	0.489
D	6	6	28.167	32.167	7	1.143	0.810
D	7	2	18.000	8.000	3	2.000	0
D	8	1	22.000	0	3	1.000	1.000
D	9	0	0	0	0	0	0
D	10	0	0	0	0	0	0

(ii) Males

Type	Year	Breeding date			No. chicks fledged		
		No.	Mean	Variance	No.	Mean	Variance
P	1	26	33.923	90.554	37	0.703	0.604
P	2	16	25.062	17.796	22	1.091	0.753
P	3	11	22.818	24.364	13	1.385	0.423
P	4	8	20.875	25.268	11	1.091	0.691
P	5	4	24.250	8.250	4	2.000	0
P	6	4	22.500	25.667	4	1.750	0.250
P	7	3	24.000	39.000	3	1.667	0.333
P	8	2	20.500	24.500	3	2.000	0
P	9	2	18.500	4.500	3	1.333	1.333
P	10	1	18.000	0	2	1.500	0.500
NOTP	1	146	29.062	56.803	211	0.948	0.688
NOTP	2	94	24.383	32.949	116	1.310	0.581
NOTP	3	58	22.845	27.572	73	1.315	0.413
NOTP	4	41	20.610	19.094	46	1.435	0.473
NOTP	5	24	19.167	20.667	30	1.400	0.524
NOTP	6	19	18.105	15.544	20	1.550	0.471
NOTP	7	9	18.222	6.944	12	1.750	0.386
NOTP	8	5	16.600	9.300	7	1.429	0.619
NOTP	9	2	12.500	4.500	3	1.667	0.333
NOTP	10	1	16.000	0	1	1.000	0
I	1	57	30.614	59.170	90	0.811	0.672
I	2	36	26.611	36.359	48	1.083	0.674
I	3	26	23.615	19.286	29	1.310	0.436
I	4	15	22.933	24.352	17	1.235	0.691
I	5	10	21.200	32.622	12	1.417	0.811
I	6	7	18.143	6.143	7	2.000	0
I	7	3	18.333	8.333	4	1.750	0.250
I	8	2	19.000	18.000	3	1.667	0.333
I	9	0	0	0	1	1.000	0
I	10	1	16.000	0	1	1.000	0
I & DI	1	92	29.707	62.034	134	0.873	0.668
I & DI	2	54	25.463	36.442	68	1.279	0.622
I & DI	3	35	23.371	24.711	40	1.300	0.421
I & DI	4	23	21.739	23.202	26	1.346	0.555
I & DI	5	14	20.286	25.912	17	1.294	0.721
I & DI	6	9	17.556	6.028	9	1.778	0.194
I & DI	7	5	17.400	5.800	6	1.833	0.167
I & DI	8	3	17.667	14.333	5	1.400	0.800
I & DI	9	0	0	0	1	1.000	0
I & DI	10	1	16.000	0	1	1.000	0
DI	1	30	28.700	64.631	37	1.027	0.694
DI	2	14	24.571	29.956	15	1.800	0.171
DI	3	6	24.667	54.267	7	1.000	0.333
DI	4	5	20.800	16.700	6	1.333	0.267
DI	5	3	18.667	4.333	4	1.000	0.667
DI	6	1	15.000	0	1	1.000	0

(ii) Males

Type	Year	Breeding date			No. chicks fledged		
		No.	Mean	Variance	No.	Mean	Variance
DI	7	1	16.000	0	1	2.000	0
DI	8	0	0	0	1	0	0
DI	9	0	0	0	0	0	0
DI	10	0	0	0	0	0	0
DI & D	1	70	28.471	53.180	96	1.073	0.679
DI & D	2	42	22.929	23.044	50	1.520	0.459
DI & D	3	22	22.136	28.314	30	1.300	0.355
DI & D	4	20	19.500	12.158	22	1.545	0.260
DI & D	5	12	18.083	8.992	15	1.467	0.410
DI & D	6	10	18.000	25.556	11	1.273	0.618
DI & D	7	4	18.500	11.000	6	1.667	0.667
DI & D	8	2	15.000	0	3	1.000	1.000
DI & D	9	2	12.500	4.500	2	2.000	0
DI & D	10	0	0	0	0	0	0
D	1	39	27.974	42.815	56	1.107	0.679
D	2	28	22.107	18.470	33	1.455	0.506
D	3	15	20.867	17.552	22	1.364	0.338
D	4	14	19.000	11·692	15	1.667	0.238
D	5	8	17.625	11.982	10	1.700	0.233
D	6	8	17.750	27.929	9	1.222	0.694
D	7	2	17.500	4.500	4	1.500	1.000
D	8	2	15.000	0	2	1.500	0.500
D	9	2	12.500	4.500	2	2.000	0
D	10	0	0	0	0	0	0

In this table, the data are of individual birds from each pair that nested in the colony on Fair Isle. The following symbols are used: P for pale, NOTP for not pale (or melanic), I for intermediate, DI for dark intermediate, and D for dark. The year refers to the number of years that a particular pair bred together in the colony; no pairs survived together for longer than 10 years.

Analysis of variance of a $2 \times r$ table with unequal numbers of observations

When data are classified in two different ways, for example breeding dates of Arctic Skuas classified by both plumage phenotype and sex, the mean values in each class form an $r \times s$ table. A classification into two sexes and three phenotypes (pale, intermediate and dark) gives rise to a 2×3 table of the means of males and females of each phenotype. Experimental work is usually designed so that equal numbers of observations are obtained in each class. Analysis of the components of variation then produces independent sums of squares for all of the $rs - 1$ degrees of freedom. Consider the $2 \times r$ table (Table B1) showing the means, numbers of observations and expectations in each class. The parameters $\alpha_1, \ldots \alpha_j,$ $\ldots \alpha_r$ measure the effects of the attributes $A_1, \ldots A_j, \ldots A_r$ on the general mean μ; β_1 and β_2 measure the corresponding effects of B_1 and B_2. If the numbers had been equal,

$$n_{11} = n_{12} = \ldots n_{1j} = n_{2j} = \ldots n_{1r} = n_{2r}$$

the effects of B_1 and B_2 would have contributed equally to the means of the A values. If so, then

$$\varepsilon(\bar{x}_{.j}) = \mu + \alpha_j + \frac{\beta_1 + \beta_2}{2} \qquad (j = 1, \ldots r)$$

The term $(\beta_1 + \beta_2)/2$ is common to all these expected values. Differences between β_1 and β_2 do not contribute to the sum of squares derived from the means of the A values: only the differences between the α values and the residual error determine the sum of squares. Similarly the sum of squares for the B values is independent of the sum of squares for the A values. The sum of squares for interaction is also independent of the other sums of squares and can easily be obtained by subtraction.

The table shows that, when unequal numbers are observed in each class,

Table B1. *Means, numbers of values and expectations in a* $2 \times r$ *table*

Effect of B	Effect of A		
	A_1	A_j	A_r
B_1	$\bar{x}_{11}, n_{11}$	$\bar{x}_{1j}, n_{1j}$	$\bar{x}_{1r}, n_{1r}$
	$\mu + \alpha_1 + \beta_1$	$\mu + \alpha_j + \beta_1$	$\mu + \alpha_r + \beta_1$
B_2	$\bar{x}_{21}, n_{21}$	$\bar{x}_{2j}, n_{2j}$	$\bar{x}_{2r}, n_{2r}$
	$\mu + \alpha_1 + \beta_2$	$\mu + \alpha_j + \beta_2$	$\mu + \alpha_r + \beta_2$
Average over B_1 and B_2	$\bar{x}._1, n_{11} + n_{21}$	$\bar{x}._j, n_{1j} + n_{2j}$	$\bar{x}._r, n_{1r} + n_{2r}$
	$\mu + \alpha_1 + \dfrac{n_{11}\beta_1 + n_{21}\beta_2}{n_{11} + n_{21}}$	$\mu + \alpha_j + \dfrac{n_{1j}\beta_1 + n_{2j}\beta_2}{n_{1j} + n_{2j}}$	$\mu + \alpha_r + \dfrac{n_{1r}\beta_1 + n_{2r}\beta_2}{n_{1r} + n_{2r}}$

In this table, the means and numbers of values are shown on the upper lines, the expectations of the means on the lower lines. The values of the αs and βs are the effects of the As and Bs on the general mean μ.

the expectations of the As are not independent of the Bs. The general term

$$(n_{1j}\beta_1 + n_{2j}\beta_2)/(n_{1j} + n_{2j})$$

is constant only if $n_{1j} = n_{2j}$. A similar argument applies to the means of the Bs. The expectation of B_1 contains the term

$$(n_{11}\alpha_1 + \ldots n_{1j}\alpha_j + \ldots n_{1r}\alpha_r)/(n_{11} + \ldots n_{1j} + \ldots n_{1r})$$

This is equal to the corresponding term in the expression for the expectation of B_2 only if

$$n_{11} = \ldots n_{1j} = \ldots n_{1r}$$
$$n_{21} = \ldots n_{2j} = \ldots n_{2r}$$

The sums of squares of the deviations of the means of the As and Bs are independent only if all numbers of observations are equal.

In my previous analysis of the breeding dates of the Arctic Skua (O'Donald, 1980c), I had used a programme supplied for a programmable calculator that gave an approximate analysis of variance for an $r \times s$ table with unequal numbers of observations. In this programme, the sums of squares are weighted by the harmonic mean of the numbers of observations in all classes. Approximate values are then obtained for the sums of squares of the As, Bs and the interaction of $A \times B$. However, Yates (1934) gave an exact method for the analysis of an $r \times s$ table by calculating adjustments to the means. This is equivalent to the general method of fitting the parameters α and β and calculating the sums of squares corresponding to them. The method has a simple solution for the case of a $2 \times r$ table, which is the form in which the Arctic Skua data have been analysed. Consider the unweighted sums and differences of the means of the As. For the attribute A_j we have the unweighted difference and unweighted mean

$$\bar{x}_{1j}-\bar{x}_{2j}$$
$$\tfrac{1}{2}(\bar{x}_{1j}+\bar{x}_{2j})$$

with expected values

$$\varepsilon(\bar{x}_{1j}-\bar{x}_{2j}) = \beta_1-\beta_2$$
$$\varepsilon(\bar{x}_{1j}+\bar{x}_{2j})/2 = \mu+\alpha_j+\frac{\beta_1+\beta_2}{2}$$

and variances derived from residuals

$$\mathrm{var}(\bar{x}_{1j}-\bar{x}_{2j}) = \sigma^2\left(\frac{1}{n_{1j}}+\frac{1}{n_{2j}}\right)$$
$$= \sigma^2/w_j$$
$$\mathrm{var}[(\bar{x}_{1j}+\bar{x}_{2j})/2] = \sigma^2/(4w_j)$$

The quantity

$$w_j = \frac{n_{1j}.n_{2j}}{n_{1j}+n_{2j}}$$

is proportional to the inverse of the variances. It is the relative weight of each difference and mean. The overall weighted mean is given by

$$\bar{x}.. = \frac{\sum\limits_{j} w_j\,(\bar{x}_{1j}+\bar{x}_{2j})/2}{\sum\limits_{j} w_j}$$

with expectation

$$\varepsilon(\bar{x}..) = \mu+\frac{\sum\limits_{j} w_j\alpha_j}{\sum\limits_{j} w_j}+\frac{\beta_1+\beta_2}{2}$$

The difference

$$\bar{x}_{1j}-\bar{x}_{2j}$$

is independent of α, while

$$\frac{\bar{x}_{1j}+\bar{x}_{2j}}{2}-\bar{x}..$$

is independent of β.

If $\beta_1=\beta_2$, then for each A,

$$\varepsilon(x_{1j}-x_{2j}) = 0$$

The weighted mean difference for all As

$$\sum\limits_{j} w_j\,(\bar{x}_{1j}-\bar{x}_{2j})\Big/ \sum\limits_{j} w_j$$

has an expectation of zero and variance

$$1 \Big/ \sum_j (w_j/\sigma^2) = \sigma^2 \Big/ \sum_j w_j$$

Therefore

$$\chi_1^2 = \frac{\left[\sum\limits_j w_j (\bar{x}_{1j} - \bar{x}_{2j})\right]^2}{\sigma^2 \sum\limits_j w_j}$$

since the right-hand side of the equation is the square of a standardized variate and hence equal to χ^2 with one degree of freedom. This holds only for the null hypothesis that $\beta_1 = \beta_2$.

On the null hypothesis that

$$\alpha_1 = \ldots \alpha_j = \ldots \alpha_r$$

we note that

$$\frac{\bar{x}_{1j} + \bar{x}_{2j}}{2} - \bar{x} \ldots$$

then has an expectation of zero and variance $\sigma^2/(4w_j)$. Therefore,

$$\chi_{r-1}^2 = \sum \left\{ \left[\frac{\bar{x}_{1j} + \bar{x}_{2j}}{2} - \bar{x} \ldots \right]^2 \Big/ \left[\sigma^2/(4w_j) \right] \right\}$$

$$= \left\{ \sum_j w_j (\bar{x}_{1j} + \bar{x}_{2j})^2 - \left[\sum_j w_j (\bar{x}_{1j} + \bar{x}_{2j}) \right]^2 \Big/ \sum_j w_j \right\} \Big/ \sigma^2$$

The sum of squares within each class of values is given by

$$\sum_{k=1}^{n_{ij}} (x_{ijk} - \bar{x}_{ij\cdot})^2$$

Within in all classes, the total sum of squares

$$\sum_{i=1}^{2} \sum_{j=1}^{r} \sum_{k=1}^{n_{ij}} (x_{ijk} - \bar{x}_{ij\cdot})^2$$

has $N - 2r$ degrees of freedom, where

$$N = \sum_i \sum_j n_{ij}$$

The effects of the As and Bs can be tested by F ratios. F is defined as the ratio

$$F_{\xi,\eta} = \frac{\chi_\xi^2/\xi}{\chi_\eta^2/\eta}$$

where ξ and η are the degrees of freedom of the two values of χ^2. The unknown parameter σ^2 is cancelled from both numerator and denominator to give a ratio of mean squares or variances. Thus the sum of squares for the

As has $r-1$ degrees of freedom. The sum of squares within classes has $N-2r$ degrees of freedom. Therefore we calculate the mean squares

M.S. (between As) = [Sum of squares (between As)]$/(r-1)$

M.S. (within classes) = [Sum of squares (within classes)]$/(N-2r)$

and hence

$$F_{r-1,\ N-2r} = \frac{\text{M.S. (between } A\text{s)}}{\text{M.S. (within classes)}}$$

So far, we have neglected the possibility of interactions between the attributes A and B, giving rise to terms $(\alpha\beta)_{ij}$ in the expectations of the means. If no interaction is present, the expression

$$\bar{x}_{1j}-\bar{x}_{2j}-\sum_j w_j\,(\bar{x}_{1j}-\bar{x}_{2j})\Big/\sum_j w_j$$

measures the deviation of each difference from its weighted mean value. This has an expectation of zero and a variance of σ^2/w_j so that

$$\chi^2_{r-1} = \sum_j\{[\bar{x}_{1j}-\bar{x}_{2j}-\sum_j w_j\,(\bar{x}_{1j}-\bar{x}_{2j})\Big/\sum_j w_j]^2\Big/[\sigma^2/w_j]\}$$

$$= \{\sum_j w_j\,(\bar{x}_{1j}-\bar{x}_{2j})^2-[\sum_j w_j\,(\bar{x}_{1j}-\bar{x}_{2j})]^2\Big/\sum_j w_j\}\Big/\sigma^2$$

If the interaction is present, it will invalidate the calculation we have used to obtain the sum of squares for the Bs: the difference

$$\bar{x}_{1j}-\bar{x}_{2j}$$

would contain the expression

$$(\alpha\beta)_{1j}-(\alpha\beta)_{2j}$$

and these expressions do not disappear from the weighted sum of the differences. The sum of squares for the Bs would not be independent of the interactions. To calculate a sum of squares for the Bs when A and B interact, we must use an expression similar to that for the As. The weights are given by

$$v_i = 1/\sum_j\,(1/n_{ij})$$

where i takes the values 1 and 2 for the effects of B_1 and B_2. Then the sum of squares for the Bs is equal to the expression

$$\sum_i v_i\,\Big(\sum_j \bar{x}_{ij}\Big)^2-\Big(\sum_i v_i \sum_j \bar{x}_{ij}\Big)^2\Big/\sum_i v_i$$

Using this general expression, we thus obtain the analysis of variance of the effects of the As, Bs and their interactions as shown in table B2.

Table B2. *Analysis of variance of a 2 × r table with unequal numbers of observations*

Source of variation	Sum of squares	Degrees of freedom	Mean square
Between effects of As	$SS(A) = \sum_j w_j \left[\sum_i \bar{x}_{ij}\right]^2 - \left[\sum_j w_j \sum_i \bar{x}_{ij}\right]^2 \Big/ \sum_j w_j$	$r-1$	$SS(A)/(r-1)$
Between effects of Bs	$SS(B) = \sum_i v_i \left[\sum_j \bar{x}_{ij}\right]^2 - \left[\sum_i \sum_j \bar{x}_{ij}\right]^2 \Big/ \sum_i v_i$	1	$SS(B)$
Interaction of As and Bs	$SS(A \times B) = \sum_j w_j (\bar{x}_{1j} - \bar{x}_{2j})^2 - \left[\sum_j w_j (\bar{x}_{1j} - \bar{x}_{2j})\right]^2 \Big/ \sum_j w_j$	$r-1$	$SS(A \times B)/(r-1)$
Within classes	$SS(W) = \sum_i \sum_j \sum_k x_{ijk}^2 - \sum_i \sum_j n_{ij}\bar{x}_{ij}^2$	$N-2r$	$SS(W)/(N-2r)$

In this table $w_j = 1/(\dfrac{1}{n_{1j}} + \dfrac{1}{n_{2j}})$, $v_i = 1/\sum_j(\dfrac{1}{n_{ij}})$ and $N = \sum_i \sum_j n_{ij}$.

The subscript i takes values 1 and 2; j takes values 1 to r.

Statistical analysis of assortative and disassortative mating in polymorphic birds

C.1 Introduction to the theory of assortative mating

Assortative mating occurs when one or more phenotypes mate more often than at random with others like themselves. It is the tendency of like to mate with like. Disassortative mating, or negative assortative mating, is the tendency of unlike phenotypes to mate with each other.

Assortative mating has often been observed between phenotypes in polymorphic populations. It evolves as sub-populations become ethologically isolated. It can evolve without the isolation or sub-division of populations if mating preferences are selected in favour of particular phenotypes: the genes for the preferences increase in frequency in association with the genes for the preferred phenotypes; the association between the genes produces assortative mating (O'Donald, 1980a). Behavioural mechanisms such as imprinting on parental phenotypes produce assortative mating (Cooke, 1978). We should expect to observe and do observe, that assortative mating takes place in many polymorphic populations.

Disassortative mating has also been observed, though it does not seem to follow from simple evolutionary or behavioural mechanisms. The obvious example of disassortative mating is sex itself, which has still to be explained satisfactorily in terms of selective advantage (Maynard-Smith, 1978). Self-incompatibility systems are other examples. Disassortative mating between the sexes is of course complete: all fertile matings are males × females. Apart from these special mechanisms preventing asexual reproduction or inbreeding, a partial tendency to disassortment in matings between different phenotypes has been observed in certain polymorphic populations of birds. In a particular urban population of pigeons, disassortative matings were found to predominate between various melanic phenotypes and the wild-type (Murton *et al.*, 1973). In one colony of Eleonora's Falcon, melanic and non-melanic phenotypes mate disassortatively, although in another colony they do not (Walter, 1979). Melanism in birds is also the subject of assortative mating: melanic Arctic Skuas mate assortatively (Davis & O'Donald, 1976a; O'Donald, 1980c); melanic and non-melanic Snow Geese mate assortatively as a result of imprinting (Cooke, 1978).

The usual method of analysing assortative and disassortative mating is to calculate the numbers of matings that would occur by purely random selection of mates and hence calculate χ^2 for the deviations of observed numbers from the expected numbers of random matings. This is usually satisfactory as far as it goes. Difficulties arise if samples of numbers of matings have been obtained from different populations. The degree of assortment in the matings and the proportions of the phenotypes may differ between populations. To test the significance of these differences, models of assortative mating must be fitted to the data and the parameters estimated. The overall value of χ^2 can then be analysed into components corresponding to the effects of assortative mating, the variation in phenotypic frequency and the residual heterogeneity that tests the goodness of fit of the model.

Davis & O'Donald (1976a) and O'Donald (1980c) analysed the assortative mating of Arctic Skuas in different Shetland colonies. They fitted models and estimated assortative mating parameters. Davis & O'Donald (1976b) fitted a specific model of disassortative mating to Murton, Westwood & Thearle's data of matings of pigeons (Murton *et al.*, 1973). Feral pigeons are highly polymorphic with at least three distinct melanic phenotypes determined by dominant genes. In the model it was assumed that the wild-type and two of the melanic phenotypes mated disassortatively. Estimation of parameters showed that about 50 per cent of females must have expressed disassortative mating preferences to give rise to the observed numbers of matings.

In this Appendix, I give the general formulae for the maximum likelihood (M.L.) estimates of mating preferences expressed assortatively or disassortatively. I use these formulae in general methods of analysis of data of both assortative and disassortative mating. To illustrate the methods of analysis, I have taken data of assortative and disassortative mating of melanic and non-melanic phenotypes of Arctic Skuas, feral pigeons and Eleonora's Falcons. Analyses of the Arctic Skua data have already been published (Davis & O'Donald, 1976a; O'Donald, 1980c). The disassortative mating of feral pigeons is analysed in more detail and the disassortative mating of Eleonora's Falcons is analysed here for the first time.

C. 2 Models of assortative and disassortative mating

In setting up the models of assortative and disassortative mating, it is assumed that the matings are produced by individuals taken from a large population. The hypothetical frequencies of the matings are determined by the population frequencies of the phenotypes and the proportions of females expressing preferences. In fitting the models, the hypothetical frequencies of matings are considered to be the probabilities of observing the matings: the sample of observed matings then follows a multinomial distribution with these hypothetical probabilities. In an experiment, a fixed number of individuals might be placed together to mate. If each pair were removed after mating, the mated individuals would then represent a hypergeometrically distributed sample taken from the finite population of the number of individuals used in the experiment. In a sample of mated individuals taken from a natural population, the size of the population will almost always be unknown. But this does not create a difficulty. The matings may be considered to occur as individuals are taken from the hypothetical infinite population of zygotes which the parents could potentially produce. The sample of matings, like a sample of individuals or genes, is then multinomially distributed with expectations determined by the actual parental frequencies. The bias in measuring sexual selection when mating pairs are chosen from a known finite population (see the critique of measuring selection by Goux & Anxolabehere, 1980) should present no problem in measuring mating preferences in samples from large natural populations.

The model of preferential mating to be fitted to the data simply assumes that some females will have preferences to mate only with certain phenotypes of males. The expression of these preferences may be determined by the females' phenotypes. In assortative mating, a preference is only expressed if the female possesses the phenotype she prefers. In disassortative mating, she only expresses her preference if she does not possess that phenotype. The theory is developed in terms of female preference, because if choices are expressed, it is almost always the females that express them. But, without loss of generality, it applies equally to male choice, or choice by both sexes. Table C1 (i) shows the matings that take place with assortative and disassortative expression of preference in favour of two distinct phenotypes. Table C1 (ii) shows the overall frequencies of matings, both preferential and random. In obtaining these overall frequencies, monogamy has been assumed. Matings could also be polygamous. In polygamy, males who have once been mated can mate again, the same male taking part in several matings. Then the frequencies of the males

Table C1. *Frequencies of matings with and without assortment of phenotypes*

(i) Frequencies of preferential matings

Frequencies of phenotypes of males mated preferentially

Female phenotypes	Without assortment		With positive assortment		With negative assortment	
	A	B	A	B	A	B
A	αu	βu	αu	—	—	βu
B	αv	βv	—	βv	αv	—

(ii) Total frequencies of both preferential and random matings

Assortment and female phenotypes		Frequencies of phenotypes of males	
		A	B
With positive assortment	A	$\alpha u + \dfrac{u^2(1-\alpha)^2}{1-\alpha u - \beta v}$	$\dfrac{uv(1-\alpha)(1-\beta)}{1-\alpha u - \beta v}$
	B	$\dfrac{uv(1-\alpha)(1-\beta)}{1-\alpha u - \beta v}$	$\beta v + \dfrac{v^2(1-\beta)^2}{1-\alpha u - \beta v}$
With negative assortment	A	$\dfrac{u(u-\alpha v)(1-\beta)}{1-\alpha v - \beta u}$	$\dfrac{uv(1-\alpha\beta)}{1-\alpha v - \beta u}$
	B	$\dfrac{uv(1-\alpha\beta)}{1-\alpha v - \beta u}$	$\dfrac{v(v-\beta u)(1-\alpha)}{1-\alpha v - \beta u}$

Note that the frequencies of reciprocal matings A female $\times B$ male and A male $\times B$ female are identical in both models.

in mating may differ from their frequencies as individuals in the population. The overall frequencies of both polygamous and monogamous matings are shown in table 3.1 of O'Donald (1980*a*). The M.L. estimates of parameters of polygamous assortative mating were also given by Davis & O'Donald (1976*a*). But only monogamous mating is considered here. Usually only one set of matings can be observed, each individual being counted once in one mating. If successive sets of matings from the same population were observed, some individuals might be counted in more than one mating in a polygamous species and a model with polygamy would then be appropriate.

Given only two phenotypes as in table C1, the data will consist of just three different types of mating:

$A \times A$, $A \times B$ and $B \times B$

Only two independent parameters can be fitted to such data: one is the frequency of A; the other is the proportion of females with a preference. Together with the total number of matings observed in the sample, these parameters exhaust the degrees of freedom. If other samples have been obtained, the additional degrees of freedom could be used to estimate additional parameters. But since one preference parameter is sufficient to fit the data when summed for all samples, additional parameters would be superfluous on the hypothesis that the samples have all been drawn from populations with the same mating preference. Additional parameters are only worth fitting when more than two phenotypes mate assortatively or disassortatively. A model with the two parameters, α and β, gives rise to three distinct one-parameter models: $\alpha=0$, $\beta=0$ and $\alpha=\beta$. If $\alpha=0$, the assortative preference is B for B, the disassortative preference A for B. If $\beta=0$, the assortative preference is A for A, the disassortative preference is B for A. But these two cases, $\alpha=0$, or $\beta=0$, are equivalent. Suppose we have matings

> $A \times B$ with a observed numbers
> $A \times B$ with b observed numbers
> $B \times B$ with c observed numbers.

By interchanging a and c both models $\alpha=0$ and $\beta=0$ can be fitted by either model. Table C2 shows the probabilities of the matings and the correlations between mates. Except in the special symmetric case when $\alpha=\beta$, the correlation between the phenotypes of mating individuals is necessarily frequency-dependent. This must always be so in models of preferential mating. It must raise doubts about the biological meaning of the models of assortative mating for quantitative characters in all of which the correlation between mates is assumed to be constant.

In table C1, a distinction is made between the reciprocal matings A male $\times B$ female and A female $\times B$ male. However, the frequencies of these two types of mating are always equal. No further information can be gained by separating the matings $A \times B$ into their reciprocal types.

Table C3 shows the M.L. estimates of the parameters and the values of the elements of the information matrices

$$\begin{pmatrix} I_{11} & I_{12} \\ I_{21} & I_{22} \end{pmatrix}$$

defined by the equations

$$I_{11} = -\varepsilon\partial^2\log L/\partial u^2$$
$$I_{12} = I_{21} = -\varepsilon\partial^2 \log L/\partial\alpha\partial u$$
$$I_{22} = -\varepsilon\partial^2\log L/\partial\alpha^2$$

where L is the log likelihood (to base e for natural logarithms).

Table C2. *Probabilities of matings and correlations between mates in models of assortative and disassortative mating*

(i) Assortative mating

Mating	Observed number	Frequencies of matings		
		Case (i) $\beta = 0$	Case (ii) $\alpha = 0$	Case (iii) $\alpha = \beta$
$A \times A$	a	$\alpha u + \dfrac{u^2(1-\alpha)^2}{1-\alpha u}$	$\dfrac{u^2}{1-\beta v}$	$u^2 + \alpha uv$
$A \times B$	b	$\dfrac{2uv(1-\alpha)}{1-\alpha u}$	$\dfrac{2uv(1-\beta)}{1-\beta v}$	$2uv(1-\alpha)$
$B \times B$	c	$\dfrac{v^2}{1-\alpha u}$	$\beta v + \dfrac{v^2(1-\beta)^2}{1-\beta v}$	$v^2 + \alpha uv$
Correlation between phenotypes of mating individuals		$\rho = \dfrac{\alpha v}{1-\alpha u}$	$\rho = \dfrac{\beta u}{1-\beta v}$	$\rho = \alpha$

(ii) Disassortative mating

Mating	Observed number	Frequencies of matings		
		Case (i) $\beta = 0$	Case (ii) $\alpha = 0$	Case (iii) $\alpha = \beta$
$A \times A$	a	$\dfrac{u(u-\alpha v)}{1-\alpha v}$	$\dfrac{u^2(1-\beta)}{1-\beta u}$	$u^2 - \alpha uv$
$A \times B$	b	$\dfrac{2uv}{1-\alpha v}$	$\dfrac{2uv}{1-\beta u}$	$2uv(1+\alpha)$
$B \times B$	c	$\dfrac{v^2(1-\alpha)}{1-\alpha v}$	$\dfrac{v(v-\beta u)}{1-\beta u}$	$v^2 - \alpha uv$
Correlation between phenotypes of mating individuals		$\rho = \dfrac{-\alpha v}{1-\alpha v}$	$\rho = \dfrac{-\beta u}{1-\beta u}$	$\rho = -\alpha$

In this table u is the population frequency of A and $v (=1-u)$ is the population frequency of B. $a+b+c=n$

The variances and co-variances of the estimates of α and u are given by the equations

$$\text{var}(\hat{u}) = I_{22}/(I_{11}\,I_{22}-I_{12}{}^2)$$
$$\text{var}(\hat{\alpha}) = I_{11}/(I_{11}\,I_{22}-I_{12}{}^2)$$
$$\text{cov}(\hat{u},\hat{\alpha}) = -I_{12}/(I_{11}\,I_{22}-I_{12}{}^2)$$

Estimates and values of information are given only for cases (i) and (iii) of the models. The values for case (ii) are obtained from case (i) by writing β for α and interchanging u and v and a and c. For example, the estimate of the assortative mating preference for case (ii) is obtained by solving the quadratic equation

$$\beta^2[a\hat{v}(2\hat{v}-1)]-\beta[\hat{v}^2(c+2b+3a)-\hat{v}(2c+3b+a)+c+b]+$$
$$n\hat{v}^2-\hat{v}(2c+b)+c=0$$

The estimate of the disassortative preference for case (ii) is given by

$$\hat{\beta} = \frac{\frac{1}{4}b^2-ac}{\frac{1}{2}b(a+\frac{1}{2}b)}$$

C.3 Analysis of data

The data that may be obtained on assortative or disassortative mating in natural populations will usually consist of counts of the different types of matings. If just one sample has been obtained, the estimation of parameters leaves no degree of freedom and the model must fit exactly. The test of significance of assortative mating is then simply a test that the estimate of the preference differs significantly from zero. This is equivalent to a test of deviation from random mating. For this test we should calculate

$$\chi^2 = \frac{n(ac-\frac{1}{4}b^2)^2}{(a+\frac{1}{2}b)^2(c+\frac{1}{2}b)^2}$$

For case (iii) of the models of both assortative and disassortative mating, this is equal to

$$\chi^2 = n\hat{\alpha}^2$$

It represents a crude test of the significance of the estimated correlation between mates. Since the correlation is

$$\rho = \pm\alpha$$

and α has M.L. estimate

$$\hat{\alpha} = \pm\frac{ac-\frac{1}{4}b^2}{(a+\frac{1}{2}b)(c+\frac{1}{2}b)}$$

therefore, $\hat{\alpha}$ has variance

$$\text{var} (\hat{\alpha}) = (1-\alpha^2)^2/(n-1)$$

and hence on the null hypothesis $\alpha=0$

$$\chi^2 = \hat{\alpha}^2(n-1) \simeq n\hat{\alpha}^2$$

being the square of a standardized and approximately normal variate. This result immediately suggests a more accurate test of significance using Fisher's z transformation of the estimated correlation coefficient

$$z = \tfrac{1}{2}\{\log_e (1+\hat{\alpha}) - \log_e (1-\hat{\alpha})\}$$

with

$$\text{var} (z) = 1/(n-3)$$

Therefore

$$\chi^2 = z^2 (n-3)$$

More generally, several independent samples may have been obtained from different areas. For example, matings of Arctic Skuas have been counted in a number of Shetland islands. Suppose k such samples have been obtained. We would have the samples of

$$a_i, b_i, c_i \qquad (i = 1, \ldots, k)$$

with totals over k samples

$$a_T = \sum_{i=1}^{k} a_i$$

$$b_T = \sum_{i=1}^{k} b_i$$

$$c_T = \sum_{i=1}^{k} c_i$$

Then, for the ith sample, we have the estimate for case (iii) of the models

$$\hat{\alpha}_i = \frac{(a_i c_i - \tfrac{1}{4}b_i^2)}{(a_i + \tfrac{1}{2}b_i)(c_i + \tfrac{1}{2}b_i)}$$

From these estimates of the correlation coefficient, we can obtain the corresponding estimates of z_i $(i=1, 2, \ldots, k)$ and their weighted mean value

$$\bar{z} = \frac{\sum_{i=1}^{k} z_i (n_i - 3)}{\sum_{i=1}^{k} (n_i - 3)}$$

Table C3. *Estimates of paramters and values of information matrix*

(i) Assortative mating	Case (i): $\beta=0$	Case (iii): $a=\beta$
Estimates of parameters	$\hat{u} = (a+\tfrac{1}{2}b)/n$ $\hat{\alpha}$ is a solution of the equation $\alpha^2\,[c\hat{u}(2\hat{u}-1)]-\alpha[\hat{u}^2(a+2b+3c)-\hat{u}(2a+3b+c)$ $\quad +a+b]+n\hat{u}^2-\hat{u}(2a+b)+a=0$	$\hat{u} = (a+\tfrac{1}{2}b)/n$ $\hat{\alpha} = \dfrac{ac-\tfrac{1}{4}b^2}{(a+\tfrac{1}{2}b)(c+\tfrac{1}{2}b)}$
Values of information matrix	$I_{11} = \dfrac{n}{1-\alpha u}\left\{\dfrac{u(1-2\alpha)^2}{\alpha+u-2\alpha u}+\dfrac{(1+v-\alpha)}{u}\right.$ $\left.\qquad +\dfrac{2(1-\alpha u)}{v}-\dfrac{\alpha^2}{1-\alpha u}\right\}$ $I_{12} = \dfrac{n}{1-\alpha u}\left\{\dfrac{u}{\alpha+u-2\alpha u}-\dfrac{1}{1-\alpha u}\right\}$ $I_{22} = \dfrac{nu}{1-\alpha u}\left\{\dfrac{(1-2u)^2}{\alpha+u-2\alpha u}+\dfrac{2v}{1-\alpha}-\dfrac{u}{1-\alpha u}\right\}$	$I_{11} = \dfrac{nu(1-\alpha)^2}{u+\alpha v}+\dfrac{nv(1-\alpha)^2}{v+\alpha u}$ $\qquad +\dfrac{n(1+v-\alpha v)}{u}+\dfrac{n(1+u-\alpha u)}{v}$ $I_{12} = \dfrac{nu}{u+\alpha v}-\dfrac{nv}{v+\alpha u}$ $I_{22} = \dfrac{nuv^2}{u+\alpha v}+\dfrac{nu^2v}{v+\alpha u}+\dfrac{2nuv}{1-\alpha}$

(ii) Disassortative mating

	Case (i): $\beta = 0$	Case (iii): $\alpha = \beta$
Estimates of parameters	$\hat{u} = (a + \tfrac{1}{2}b)/n$ $\hat{\alpha} = \dfrac{\tfrac{1}{2}b^2 - 2ac}{b(c + \tfrac{1}{2}b)}$	$\hat{\alpha} = \dfrac{\tfrac{1}{4}b^2 - ac}{(a + \tfrac{1}{2}b)(c + \tfrac{1}{2}b)}$

Values of information matrix

Case (i): $\beta = 0$

$$I_{11} = \frac{n}{1-\alpha v}\left\{ \frac{1+v-\alpha v}{u} + \frac{2(1-\alpha v)}{v} \right\} + \frac{u(1+\alpha)^2}{u-\alpha v} - \frac{\alpha^2}{1-\alpha v}$$

$$I_{12} = \frac{n}{1-\alpha v}\left\{ \frac{1}{1-\alpha v} - \frac{u}{u-\alpha v} \right\}$$

$$I_{22} = \frac{mv^2}{1-\alpha u}\left\{ \frac{u}{u-\alpha v} + \frac{1}{1-\alpha} - \frac{1}{1-\alpha v} \right\}$$

Case (iii): $\alpha = \beta$

$$I_{11} = \frac{n(1+v+\alpha v)}{u} + \frac{n(1+u+\alpha u)}{v}$$
$$+ \frac{nu(1+\alpha)^2}{u-\alpha v} + \frac{nv(1+\alpha)^2}{v-\alpha u}$$

$$I_{12} = \frac{nu}{u-\alpha v} - \frac{nv}{v-\alpha u}$$

$$I_{22} = \frac{nuv^2}{u-\alpha v} + \frac{nu^2v}{v-\alpha u} + \frac{2nuv}{1+\alpha}$$

If $a_i+b_i+c_i=n_i$ then the variance of $z_i-\bar{z}$ is $1/(n_i-3)$. Therefore a test of homogeneity of the assortative or disassortative mating is given by

$$\chi^2_{k-1} = \sum_{i=1}^{k} (z_i-\bar{z})^2(n_i-3)$$

But the test does contain a bias. This bias is quite unimportant for one sample but would be cumulative over many samples. The transformed value of z is biassed upwards by the amount $\frac{1}{2}\rho/(n-1)$ which should therefore be subtracted from the z values. Obviously ρ, the hypothetical correlation, must be estimated. The null hypothesis assumes that the samples are drawn from populations with the same assortative or disassortative mating preferences. Thus the best estimate of ρ is $\bar{\alpha}$. We must therefore subtract $\frac{1}{2}\bar{\alpha}/(n-1)$ from each z_i and $\bar{z}$

C.3.1 *Analysis of χ^2 of data of mating frequencies*

In the general null hypothesis, the samples of numbers of matings are drawn from the same random mating population. In this population phenotype A occurs at frequency u and phenotype B at frequency v ($=1-u$). The matings occur at frequencies

$A \times A \quad u^2$
$A \times B \quad 2uv$
$B \times B \quad v^2$

We estimate u and v from the totals of all samples:

$\bar{u} = (a_T+\frac{1}{2}b_T)/n_T$
$\bar{v} = 1-\bar{u} = (c_T+\frac{1}{2}b_T)/n_T$

The expected numbers of matings observed in the ith sample are then:

$A \times A \quad n_i\bar{u}^2 = \varepsilon(a_i)$
$A \times B \quad 2n_i\bar{u}\,\bar{v} = \varepsilon(b_i)$
$B \times B \quad n_i\bar{v}^2 = \varepsilon(c_i)$

Hence we test this null hypothesis by calculating

$$\chi^2_{2k-1} = \sum_{i=1}^{k} \left\{ \frac{[a_i-\varepsilon(a_i)]^2}{\varepsilon(a_i)} + \frac{[b_i-\varepsilon(b_i)]^2}{\varepsilon(b_i)} + \frac{[c_i-\varepsilon(c_i)]^2}{\varepsilon(c_i)} \right\}$$

where χ^2 has $2k-1$ degrees of freedom since there are $3k$ classes of observed numbers with k degrees of freedom lost for the k values of the sample totals n_i and 1 degree of freedom lost for the fitted parameter $\bar{u}$.

There are three sources of variation that may produce a significant value of this χ^2_{2k-1}: assortative or disassortative mating, variation in values of u_i between samples, and a residual heterogeneity determined by variation between samples. Suppose we consider that disassortative mating takes place between the A phenotypes. The model for case (i) will then be appropriate. We shall estimate the average disassortative mating preference by the overall M.L. estimate

$$\bar{\alpha} = \frac{\frac{1}{2}b_T{}^2 - 2a_Tc_T}{b_T(c_T + \frac{1}{2}b_T)}$$

The expected numbers of matings observed in the ith sample will now be:

$$A \times A \quad n_i \bar{u} \, (\bar{u} - \bar{\alpha}\bar{v})/(1 - \bar{\alpha}\bar{v})$$
$$A \times B \quad 2n_i \bar{u}\bar{v}/(1 - \bar{\alpha}\bar{v})$$
$$B \times B \quad n_i \bar{v}^2 \, (1 - \bar{\alpha})/(1 - \bar{\alpha}\bar{v})$$

Thus, having incorporated disassortative mating in the model, we again calculate χ^2 as before. This time it is χ^2_{2k-2} having lost a further degree of freedom by the fitted parameter $\bar{\alpha}$. The significance of the assortative mating is given approximately by

$$\chi^2_1 = \chi^2_{2k-1} - \chi^2_{2k-2}$$

since the values of χ^2 are asymptotically additive for the fitting of each of these parameters. We can calculate a value of χ^2 for the variation in phenotypic frequency between samples. This is obtained from the $2 \times k$ contingency table of numbers of phenotypes:

A	B
$2a_1 + b_1$	$2c_1 + b_1$
.	.
.	.
.	.
$2a_i + b_i$	$2c_i + b_i$
.	.
.	.
$2a_k + b_k$	$2c_k + b_k$

Brandt and Snedecor's formula can be used to calculate χ^2_{k-1}. Thus we have now analysed χ^2 into independent, approximately additive components as shown in table C4. The variation in phenotypic frequency could also have been obtained by subtraction by calculating expected frequencies for each sample based on the overall estimate $\bar{\alpha}$ and the individual sample estimates of u:

$$\hat{u}_i = (a_i + \frac{1}{2}b_i)/n_i$$

As I have already emphasized this is only an approximate analysis of χ^2: the values of χ^2 are only asymptotically additive with increasing sample

Table C4. *Analysis of χ^2 for data of assortative or disassortative mating*

Source of variation	Value of χ^2	Degrees of freedom
Assortative or disassortative mating	$\chi^2_{2k-1} - \chi^2_{2k-2}$	1
Variation in phenotypic frequency between samples	χ^2_{k-1}	$k-1$
Residual heterogeneity	$\chi^2_{2k-2} - \chi^2_{k-1}$	$k-1$
Total	χ^2_{2k-1}	$2k-1$

size. Generally, the values of χ^2 agree closely with those obtained by other methods – for example the value of χ^2 used to test the overall significance of assortative mating given by

$$\chi^2_1 = n\bar{\alpha}^2$$

or for any model

$$\chi^2_1 = \bar{\alpha}^2/\text{var}(\bar{\alpha})$$

Different models will give small differences in the values of χ^2_1 because of the different parameterizations of the models and hence different effects of bias. With increasing sample size, the different methods of calculating χ^2 should all give convergent results. The differences are usually far too small to cause doubt in making inferences. But where differences are only just significant, it would be wise to use alternative tests.

If the phenotypic frequencies differ between samples, the general null hypothesis is refuted. If the frequencies differ widely between samples, it is not valid to obtain an overall estimate of assortative mating from the total frequencies of the matings summed over all samples. In extreme examples, this procedure can lead to an apparent assortment of the phenotypes in the matings when there is no assortment at all in any of the samples taken separately. If some samples do show significant assortative mating, an estimate of the mean overall assortative mating is given by the weighted mean

$$\bar{\alpha} = \frac{\sum\limits_{i=1}^{k} \left[\hat{\alpha}_i/\text{var}(\hat{\alpha}_i)\right]}{\sum\limits_{i=1}^{k} \left[1/\text{var}(\hat{\alpha}_i)\right]}$$

The heterogeneity in the values of $\hat{\alpha}_i$ can be tested by the χ^2

$$\chi^2_{k-1} = \sum_{i=1}^{k} \left[(\hat{\alpha}_i - \bar{\alpha})^2 / \mathrm{var}(\hat{\alpha}_i) \right]$$

This test is similar to the test of heterogeneity in values of z_i. Since the values of z_i are approximately normally distributed, the test of heterogeneity of z is more accurate. If the samples were found to differ in phenotypic frequencies, it is unclear what this test would test. The samples could not have been drawn from the same population. There would be no reason to suppose that the mating preferences might be equal in the different populations from which the samples had been drawn. The test of heterogeneity would then be a test of a meaningless null hypothesis.

C.4   Assortative and disassortative mating of melanic birds

Polymorphisms of melanic and non-melanic phenotypes can be found in many species of animals. In the Lesser Snow Goose (*Anser caerulescens*) and Arctic Skua (*Stercorarius parasiticus*), melanics mate assortatively. In the feral pigeon (*Columba livia*) and Eleonora's Falcon (*Falco eleonorae*), they mate disassortatively. The snow goose presents a special problem since reciprocal matings of white (W) × blue melanic (B) do not occur in equal numbers: W female × B male matings predominate over W male × B female matings (see table 1 of Cooke, 1978). But according to the models in table C1, the frequencies of these matings should be equal. Cooke explained the difference in frequencies by a special model of assortative mating by imprinting on parental phenotype. Simulations at certain parameter values of the model produced a close correspondence to the observed frequencies of W female × B male and W male × B female matings. The models described in this paper are not applicable to the assortative mating of the snow goose, though a generalization of the models with different frequencies in males and females could be fitted to the snow goose data.

C.4.1   Assortative mating of the Arctic Skua

Although the melanic phenotypes of the Arctic Skua can be classified as either dark homozygotes or intermediate heterozygotes, the melanics are at least semi-dominant to the pale, non-melanics. On different islands of the Shetlands, observations have been made of numbers of matings of Arctic Skuas classified simply as either melanic or non-melanic (Davis & O'Donald, 1976a; O'Donald, 1980c). If we assume that melanics mate assortatively without discrimination of intermediate or dark

Table C5. *Analysis of assortative mating of melanic and pale phenotypes of the Arctic Skua*

(i) Numbers of matings, estimates of preference and tests of significance

Matings	Fair Isle 1973–79	Foula 1975	Shetland 1946–59	Total data
Melanic × melanic	254	144	218	616
Melanic × pale	119	86	120	325
Pale × pale	19	26	38	83
Case (i):$\hat{\alpha}$	0.2159	0.3898	0.4434	0.3931*
$\chi^2 = \hat{\alpha}^2/\mathrm{var}(\hat{\alpha})$	1.417	9.801	21.932	30.021
Case (iii): $\hat{\alpha}$	0.05225	0.1469	0.1719	0.1202*
$\chi^2 = \hat{\alpha}^2(n-1)$	1.067	5.500	11.085	17.116
Value of z	0.05229	0.1479	0.1737	0.1207*
$\chi^2 = z^2(n-3)$	1.064	5.537	11.248	17.276

For test of heterogeneity of values of z:

$$\chi_2^2 = \sum_{i=1}^{3} (z_i - \bar{z})^2(n_i - 3) = 3.055$$

The correction for bias in z does not alter the last figure.

(ii) Analysis of χ^2 by fitting case (i) of the model

Component of variation	Value of χ^2	Dfs	Value of P
Preferential mating for melanics $\bar{\alpha} = 0.3826$**	19.2720	1	1.13×10^{-5}
Difference in proportions of phenotypes	11.0010	2	0.00408
Residual heterogeneity	1.4882	2	0.475
Total	31.7612	5	6.62×10^{-6}

A slightly different analysis will be obtained if case (iii) of the model is fitted.
* These estimates for the total data are the weighted means of the estimates for each sample.
** This estimate is derived from the total data of the numbers of matings summed over all samples. It differs only slightly from the weighted mean.

melanics, then case (i) of the model applies to the data. Case (iii) of the model applies if both melanics and pales mate assortatively. The data, the estimates of the assortative mating parameters, and the analysis of χ^2 are shown in table C5. The assortative mating is very highly significant, with no significant heterogeneity in the estimates obtained from different colonies.

The test of departure from random mating is given by

$$\chi_1^2 = \hat{\alpha}^2 (n-1)$$

where $\hat{\alpha}$ is the estimate of the correlation between mates obtained from case (iii) of the model. The use of the transformed estimate of correlation, z, hardly alters the values of χ^2. On the other hand, when the formula

$$\chi^2 = \hat{\alpha}^2/\text{var}(\hat{\alpha})$$

is applied to the estimate obtained from case (i) of the model, values of χ^2 are seriously overestimated. No doubt, these errors arise because this estimate of α is far from being normally distributed.

An obvious discrepancy can be seen in the values of χ^2 for the test of heterogeneity of z and for the residual heterogeneity in the analysis of χ^2. Neither value has any statistical significance: the inference is the same. The discrepancy occurs because χ^2 is only asymptotically additive when different components of the model are fitted. Usually, closer agreement is obtained between the different methods of analysis. For example if case (iii) of the model is fitted to the data, we get the following analysis:

Component of variation	Value of χ^2	Dfs
Mating preference	17.7119	1
Difference in proportions of phenotypes	11.0010	2
Residual heterogeneity	3.0483	2

In this analysis, the value of χ^2 for residual heterogeneity is very close to that for heterogeneity between values of z: the χ^2 for preferential mating is close to that for the significance of z. Exact additivity of χ^2 holds only for independent linear functions of the observations.

If the melanic phenotypes are also classified as either dark or intermediate, a model with three parameters – preference for dark, preference for intermediate and preference for pale – can be fitted to the data. Davis & O'Donald (1976a) obtained data classified in this way from Arctic Skua colonies on the islands of Fair Isle and Foula. They fitted a model with the three parameters to their data. Explicit formulae for the M.L. estimates cannot be found, however. A trial and error method was therefore used to find the values of the parameters that maximized the likelihood. The results are shown in table 4 of Davis & O'Donald's paper.

C.4.2 Disassortative mating of the feral pigeon and Eleonora's Falcon

Murton et al. (1973) observed the following matings in an urban population of feral pigeons:

Matings	$C^+ \times C^+$	$C^+ \times C$	$C^+ \times C^T$	$C^+ \times S$	$C \times C$
Numbers	2	48	35	12	5
Matings	$C \times C^T$	$C \times S$	$C^T \times C^T$	$C^T \times S$	$S \times S$
Numbers	57	18	3	13	2

Table C6. *Disassortative matings of each phenotype in a population of feral pigeons*

Phenotype	Wild-type	Blue checker	T-pattern	Others
Matings	$C^+ \times C^+$ 2	$C \times C$ 5	$C^T \times C^T$ 3	$S \times S$ 2
and	$C^+ \times \tilde{C}^+$ 95	$C \times \tilde{C}$ 123	$C^T \times \tilde{C}^T$ 105	$S \times \tilde{S}$ 43
numbers	$\tilde{C}^+ \times \tilde{C}^+$ 98	$\tilde{C} \times \tilde{C}$ 67	$\tilde{C}^T \times \tilde{C}^T$ 87	$\tilde{S} \times \tilde{S}$ 150
Preferences				
Case (i)	0.2981	0.4362	0.3407	0.0440
Case (iii)	0.2861	0.4034	0.3223	0.0403
Values of z	0.2943	0.4277	0.3342	0.0403
$\chi^2 = z^2 (n-3)$	16.626	35.125	21.445	0.312

The phenotypes are as follows: C^+ is the non-melanic wild-type, C and C^T are the melanics blue-checker and T-pattern; S consists of spread pattern and other rare phenotypes. A glance at the numbers is sufficient to show that most of the matings are disassortative. Table C6 shows the data classified into matings of one phenotype with all other phenotypes lumped together in a single class, for example matings of wild-type (C^+) and not wild-type ($\tilde{C}^+$). Wild-types, blue checkers and T-patterns all mate disassortatively. The other phenotypes show no evidence of assortative or disassortative mating. Davis & O'Donald (1976*b*) fitted a specific model to these data assuming that C^+, C and C^T would each express disassortative preferences for each of these three phenotypes. Disassortative preferences for C and C^T were found sufficient to fit the data, leaving no significant residual heterogeneity.

As part of a very detailed and extensive study of the ecology of Eleonora's Falcon, Walter (1979) observed the frequencies of matings of melanic and non-melanic phenotypes of the birds in two separate island colonies. Walter made no analysis of his data, which are shown in table C7 with my analysis of χ^2 corresponding to that for the assortative mating of the Arctic Skua. The values of χ^2 are all in close agreement. The heterogeneity in values of z is tested by

$$\chi_1^2 = 7.891$$

while the residual heterogeneity in the analysis of χ^2 is tested by

$$\chi_1^2 = 7.914$$

In the colony of falcons on Paximada, the disassortative mating is statistically very significant. On Mogador, melanics appear to mate assortatively, not disassortatively, but this is not significant. The disassor-

Table C7. *Disassortative mating of Eleonora's Falcon*

(i) Numbers of matings, estimates of
preference and tests of significance

Matings	Paximada	Mogador
Light × light	32	44
Light × dark	45	19
Dark × dark	3	5
Case (i) $\hat{\alpha}$	0.3346	−0.2553
Case (iii) $\hat{\alpha}$	0.2952	−0.1673
Value of z	0.3042	−0.1688
$\chi^2 = z^2(n-3)$	7.128	1.853

For test of heterogeneity of values of z
$$\chi_1^2 = 7.891$$

(ii) Analysis of χ^2 by fitting case (i) of the model

Component of variation	Value of χ^2	Dfs	Value of P
Preferential mating for melanics $\hat{\alpha}=0.1204$	1.6430	1	0.200
Difference in proportions of phenotypes	4.1499	1	0.0416
Residual heterogeneity	7.9137	1	0.00491
Total	13.7066	3	0.00333

tative mating system on Paximada differs very significantly from the mating system on Mogador. This is a remarkable finding. If it had been only just significant, it would have been tempting to discount it as a freak result. But the difference observed would only arise by random sampling with the probability $P=0.0051$. It might possibly have arisen by random sampling when the colonies were originally founded. They would probably have been founded by a few pairs of birds. The original founders may have differed by chance in their mating preferences. If the young birds usually return to breed in the colony where they were born, the original difference due to sampling would have persisted as the numbers increased, thus giving rise to the present difference between the mating systems of the two colonies. An hypothesis such as this, that depends on the occurrence of an untestable historical event, is clearly unsatisfactory, but I cannot give any alternative explanation.

References

Andersson, M. (1976). Predation and kleptoparasitism by skuas in a Shetland seabird colony. *Ibis*, **118**, 208–17.

Andersson, M. & Götmark, F. (1980). Social organization and foraging ecology in the Arctic Skua *Stercorarius parasiticus*: a test of the food defendability hypothesis. *Oikos*, **35**, 63–71.

Arnason, E. (1978). Apostatic selection and kleptoparasitism in the Parasitic Jaeger. *The Auk*, **95**, 377–81.

Arnason, E. & Grant, P. R. (1978). The significance of kleptoparasitism during the breeding season in a colony of Arctic Skuas *Stercorarius parasiticus* in Iceland. *Ibis*, **120**, 38–54.

Barlow, G. W. (1974). Hexagonal territories. *Animal Behaviour*, **22**, 876–8.

Barrington, R. M. (1890). The Great Skua in Foula. *Zoologist*, 297–301.

Baxter, E. V. & Rintoul, L. J. (1953). *The Birds of Scotland*. Edinburgh: Oliver and Boyd.

Bengtson, S. A. & Owen, D. F. (1973). Polymorphism in the Arctic Skua *Stercorarius parasiticus* in Iceland. *Ibis*, **115**, 87–92.

Berry, R. J. & Davis, P. E. (1970). Polymorphism and behaviour in the Arctic Skua (*Stercorarius parasiticus* (L.)). *Proceedings of the Royal Society of London*, **B175**, 255–67.

Bishop, J. A. (1972). An experimental study of the cline of industrial melanism in *Biston betularia* (L.) (Lepidoptera) between urban Liverpool and rural North Wales. *Journal of Animal Ecology*, **41**, 209–43.

Bourne, W. R. P. & Dixon, T. J. (1974). The seabirds of the Shetlands. *Seabird Reports*, **4**, 1–18.

Brown, J. L. (1964). The evolution of diversity in avian territorial systems. *Wilson Bulletin*, **76**, 160–9.

Bulmer, M. G. & Perrins, C. M. (1973). Mortality in the Great Tit *Parus major*. *Ibis*, **115**, 277–81.

Burger, J. (1980). Territory size differences in relation to reproductive stage and type of intruder in Herring Gulls (*Larus argentatus*). *The Auk*, **97**, 733–41.

Charlesworth, D. & Charlesworth, B. (1975). Sexual selection and polymorphism. *American Naturalist*, **109**, 465–70.

Clarke, B. (1962). Natural selection in mixed populations of two polymorphic snails. *Heredity*, **17**, 319–45.

Clarke, B. & O'Donald, P. (1964). Frequency-dependent selection. *Heredity*, **19**, 201–6.

Clarke, C. A. & Sheppard, P. M. (1966). A local survey of the distribution of industrial melanic forms in the moth *Biston betularia* and estimates of the selective values of these in an industrial environment. *Proceedings of the Royal Society of London*, **B165**, 424–39.

Clutton-Brock, T. H., Albon, S. T., Gibson, R. M. & Guinness, F. E. (1979). The logical stag: adaptive aspects of fighting in Red Deer (*Cervus alephus*, L.). *Animal Behaviour*, **27**, 211–25.

Clutton-Brock, T. H., Harvey, P. H. & Rudder, B. (1977). Sexual dimorphism, socionomic sex ratio and body weight in primates. *Nature*, **269**, 797–800.

Cooke, F. (1978). Early learning and its effect on population structure. Studies of a wild population of Snow Geese. *Zeitschrift für Tierpsychologie*, **46**, 344–58.

Cooke, F. & Cooch, F. G. (1968). The genetics of polymorphism in the Goose *Anser caerulescens*. *Evolution*, **22**, 289–300.

Cooke, F., Mirsky, P. J. & Seiger, M. B. (1972). Color preferences in the Lesser Snow Goose and their possible role in mate selection. *Canadian Journal of Zoology*, **50**, 529–36.

Cramp, S., Bourne, W. R. P. & Saunders, D. (1974). *The Seabirds of Britain and Ireland*. London: Collins.

Crook, J. H. (1965). The adaptive significance of avian social organizations. *Symposia of the Zoological Society of London*, **14**, 181–218.

Darwin, C. R. (1859). *On the Origin of Species by Means of Natural Selection, or the Preservation of Favoured Races in the Struggle for Life.* London: John Murray.

Darwin, C. R. (1871). *The Descent of Man, and Selection in Relation to Sex.* London: John Murray.

Davis, J. W. F. & O'Donald, P. (1976a). Estimation of assortative mating preferences in the Arctic Skua. *Heredity*, **36**, 235–44.

Davis, J. W. F. & O'Donald, P. (1976b). Territory size, breeding time and mating preference in the Arctic Skua. *Nature*, **260**, 774–5.

Dawkins, R. (1976). *The Selfish Gene.* Oxford: Oxford University Press.

Dean, F. C., Valkenburg, P. & Magoun, A. J. (1976) Inland migration of jaegers in Northeastern Alaska. *Condor*, **78**, 271–3.

Dott, H. E. M. (1967). Number of Great Skuas and other seabirds of Hermaness, Unst. *Scottish Birds*, **4**, 340–50.

Dublin, L. I. & Lotka, A. J. (1925). On the true rate of natural increase as exemplified by the population of the United States, 1920. *Journal of the American Statistical Association*, **20**, 305–39.

Edwards, A. W. F. (1972). *Likelihood.* Cambridge: Cambridge University Press.

Ehrman, L. (1968). Frequency-dependence of mating success in *Drosophila pseudoobscura*. *Genetical Research*, **11**, 134–40.

Ehrman, L. (1972). Genetics and sexual selection. In *Sexual Selection and the Descent of Man 1871–1971*, ed. B. Campbell, pp. 105–35. London: Heinemann.

Ehrman, L. & Spiess, E. B. (1969). Rare type mating advantage in *Drosophila*. *American Naturalist*, **103**, 675–80.

Fabritius, H. E. (1969). Notes on the birds of Foula. *Ardea*, **57**, 158–62.

Falconer, D. S. (1960). *Introduction to Quantitative Genetics.* Edinburgh: Oliver and Boyd.

Falconer, D. S. (1980). *Introduction to Quantitative Genetics*, 2nd edn. London: Longman Group Ltd.

Farr, J. A. (1980). Social behaviour patterns as determinants of reproductive success in the Guppy, *Poecilia reticulata* Peters (Pisces: Poeciliidae). An experimental study of the effects of intermale competition, female choice and sexual selection. *Behaviour*, **74**, 38–91.

Fisher, R. A. (1930). *The Genetical Theory of Natural Selection*. Oxford: Clarendon Press.

Fisher, R. A. (1937). The wave of advance of advantageous genes. *Annals of Eugenics*, **7**, 355–69.

Fisher, R. A. (1950). Gene frequencies in a cline determined by selection and diffusion. *Biometrics*, **6**, 353–61.

Furness, R. W. (1977). Effects of Great Skuas on Arctic Skuas in Shetland. *British Birds*, **70**, 96–107.

Furness, R. W. (1978). Kleptoparasitism by Great Skuas (*Catharacta skua* Brünn.) and Arctic Skuas (*Stercorarius parasiticus* L.) at a Shetland seabird colony. *Animal Behaviour*, **26**, 1167–77.

Furness, B. L. & Furness, R. W. (1980). Apostatic selection and kleptoparasitism in the Parasitic Jaeger: a comment. *The Auk*, **97**, 832–6.

Goux, J. M. & Anxolabehere, D. (1980). The measurement of sexual isolation and selection: a critique. *Heredity*, **45**, 255–62.

Grant, P. R. (1968). Polyhedral territories of animals. *American Naturalist*, **102**, 75–80.

Grant, P. R. (1971). Interactive behaviour of Puffins (*Fratercula arctica* L.) and Skuas (*Stercorarius parasiticus* L.). *Behaviour*, **40**, 263–81.

Grant, B., Snyder, G. A. & Glessner, S. F. (1974). Frequency-dependent male selection in *Mormoniella vitripennis*. *Evolution*, **28**, 259–64.

Haldane, J. B. S. (1948). The theory of a cline. *Journal of Genetics*, **48**, 277–84.

Harris, M. P. (1976). The seabirds of Shetland in 1974. *Scottish Birds*, **9**, 37–68.

Hasegawa, M. & Tanemura, M. (1976). On the pattern of space division by territories. *Institute of Statistical Mathematics, Research Memorandum* No. 89.

Hilden, O. (1971). Occurrence, migration and colour phases of the Arctic Skua (*Stercorarius parasiticus*) in Finland. *Annales Zoologici Fennici*, **8**, 223–30.

Huxley, J. S. (1938a). The present standing of the theory of sexual selection. In *Evolution: Essays on Aspects of Evolutionary Biology*, ed. G. R. de Beer, pp. 11–42. Oxford: Clarendon Press.

Huxley, J. S. (1938b). Darwin's theory of sexual selection and the data subsumed by it, in the light of recent research. *American Naturalist*, **72**, 416–33.

Jackson, E. E. (1966). Birds of Foula. *Scottish Birds*, **4**, 1–60.

Karlin, S. & O'Donald, P. (1978). Some population genetic models combining sexual selection with assortative mating. *Heredity*, **41**, 165–74.

Karlin, S. & Raper, J. (1979). Sexual selection encounter models. *Theoretical Population Biology*, **15**, 246–56.

Karlin, S. & Richter-Dyn, N. (1976). Some theoretical analyses of migration selection interaction in a cline: a generalized two range environment. In *Population Genetics and Ecology*, ed. S. Karlin & E. Nevo, pp. 659–706. New York: Academic Press.

Kettlewell, B. (1973). *The Evolution of Melanism: the study of a Recurring Necessity*. Oxford: Clarendon Press.

Kimball, A. W. (1954). Short-cut formulas for the exact partition of χ^2 in contingency tables. *Biometrics*, **10**, 452–8.

Lack, D. (1954). *The Natural Regulation of Animal Numbers*. Oxford: Clarendon Press.

Lack, D. (1968). *Ecological Adaptations for Breeding in Birds*. London: Chapman and Hall.

Lasiewski, R. C. & Dawson, W. R. (1967). A re-examination of the relationship between standard metabolic rate and body weight in birds. *Condor*, **69**, 13–23.

Lewontin, R. C. (1977). Caricature of Darwinism. *Nature*, **266**, 283–4.

Maxwell, A. E. (1961). *Analysing Qualitative Data*. London: Methuen and Co. Ltd.

Maynard-Smith, J. (1978). *The Evolution of Sex*. Cambridge: Cambridge University Press.

Moodie, G. E. E. (1972). Predation, natural selection and adaptation in an unusual Threespine Stickleback. *Heredity*, **28**, 155–67.

Murton, R. K. & Westwood, N. J. (1977). *Avian Breeding Cycles*. Oxford: Clarendon Press.

Murton, R. K., Westwood, N. J. & Thearle, R. J. P. (1973). Polymorphism and the evolution of a continuous breeding season in the pigeon, *Columba livia*. *Journal of Reproduction and Fertility, Supplement*, **19**, 563–77.

Nei, M. (1975). *Molecular Population Genetics and Evolution*. Amsterdam: North-Holland.

O'Donald, P. (1960*a*) Assortative mating in a population in which two alleles are segregating. *Heredity*, **15**, 389–96.

O'Donald, P. (1960*b*). Inbreeding as a result of imprinting. *Heredity*, **15**, 79–85.

O'Donald, P. (1963). Sexual selection and territorial behaviour. *Heredity*, **18**, 361–4.

O'Donald, P. (1971). Natural selection for quantitative characters. *Heredity*, **27**, 137–53.

O'Donald, P. (1972*a*). Natural selection of reproductive rates and breeding times and its effect on sexual selection. *American Naturalist*, **106**, 368–79.

O'Donald, P. (1972*b*). Sexual selection by variations in fitness at breeding time. *Nature*, **237**, 349–51.

O'Donald, P. (1973). Frequency-dependent sexual selection as a result of variations in fitness at breeding time. *Heredity*, **30**, 351–68.

O'Donald, P. (1974). Polymorphisms maintained by sexual selection in monogamous species of birds. *Heredity*, **32**, 1–10.

O'Donald, P. (1976). Mating preferences and their genetic effects in models of sexual selection for colour phases of the Arctic Skua. In *Population Genetics and Ecology*, ed. S. Karlin & E. Nevo, pp. 411–30. New York: Academic Press.

O'Donald, P. (1977). Sexual selection and the evolution of territoriality in birds. In *Lecture Notes in Biomathematics 19 Measuring Selection in Natural Populations*, ed. F. B. Christiansen & T. M. Fenchel, pp. 113–29. Berlin: Springer-Verlag.

O'Donald, P. (1978). Theoretical aspects of sexual selection: a generalized model of mating behaviour. *Theoretical Population Biology*, **13**, 226–43.

O'Donald, P. (1979). Theoretical aspects of sexual selection: variation in threshold of female mating response. *Theoretical Population Biology*, **15**, 191–204.

O'Donald, P. (1980*a*). *Genetic Models of Sexual Selection*. Cambridge: Cambridge University Press.

O'Donald, P. (1980*b*). Genetic models of sexual and natural selection in monogamous organisms. *Heredity*, **44**, 391–415.

O'Donald, P. (1980c). Sexual selection by female choice in a monogamous bird: Darwin's theory corroborated. *Heredity*, **45**, 201–17.

O'Donald, P. & Davis, J. W. F. (1975). Demography and selection in a population of Arctic Skuas. *Heredity*, **35**, 75–83.

O'Donald, P. & Davis, J. W. F. (1976). A demographic analysis of the components of selection in a population of Arctic Skuas. *Heredity*, **36**, 343–50.

O'Donald, P. & Davis, P. E. (1959). The genetics of the colour phases of the Arctic Skua. *Heredity*, **13**, 481–6.

O'Donald, P., Wedd, N. S. & Davis, J. W. F. (1974). Mating preferences and sexual selection in the Arctic Skua. *Heredity*, **33**, 1–16.

Owen, A. R. G. (1953). A genetical system admitting of two distinct stable equilibria under natural selection. *Heredity*, **7**, 97–102.

Pennie, I. D. (1948). Summer bird notes from Foula. *Scottish Naturalist*, **60**, 157–63.

Pennie, I. D. (1953). The Arctic Skua in Caithness. *British Birds*, **46**, 105–8.

Perdeck, A. C. (1963). The early reproductive behaviour of the Arctic Skua *Stercorarius parasiticus* (L.). *Ardea*, **51**, 1–15.

Perry, R. (1948). *Shetland Sanctuary*. London: Faber.

Petit, C. (1954). L'isolement sexuel chez *Drosophila melanogaster*. Etude du mutant white et de son allélomorphe sauvage. *Bulletin Biologique de la France et de la Belgique*, **88**, 435–43.

Pitt, F. (1922). The Great and Arctic Skuas in the Shetlands. *British Birds*, **16**, 174–81, 198–202.

Popper, K. R. (1972). *Objective Knowledge. An Evolutionary Approach*. Oxford: Clarendon Press.

Raper, J. K., Karlin, S. & O'Donald, P. (1979). An assortative mating encounter model. *Heredity*, **43**, 27–34.

Semler, D. E. (1971). Some aspects of adaptation in a polymorphism for breeding colours in the Threespine Stickleback (*Gasterosteus aculeatus*). *Journal of Zoology, London*, **165**, 291–302.

Southern, H. N. (1943). The two phases of *Stercorarius parasiticus* (Linnaeus). *Ibis*, **85**, 443–85.

Spiess, E. B. (1968). Low frequency advantage in mating of *Drosophila pseudoobscura* karyotypes. *American Naturalist*, **102**, 363–79.

Spiess, L. D. & Spiess, E. B. (1969). Minority advantage in interpopulational matings of *Drosophila persimilis*. *American Naturalist*, **103**, 155–72.

Steele, R. G. D. & Torrie, J. H. (1980). *Principles and Procedures of Statistics*, 2nd edn. New York: McGraw-Hill.

Taylor, I. R. (1979). The kleptoparasitic behaviour of the Arctic Skua *Stercorarius parasiticus* with three species of tern. *Ibis*, **121**, 274–82.

Trillmich, F. (1978). Feeding territories and breeding success of South Polar Skuas. *The Auk*, **95**, 23–33.

Walter, H. (1979). *Eleonora's Falcon. Adaptations to Prey and Habitat in a Social Raptor*. Chicago: University of Chicago Press.

Watson, A. (1970). Territorial and reproductive behaviour of Red Grouse. *Journal of Reproduction and Fertility, Supplement*, **11**, 3–14.

Watson, A. & Moss, R. (1971). Spacing as affected by territorial behaviour, habitat and nutrition in Red Grouse (*Lagopus lagopus scoticus*). In *Behaviour and Environment: the Use of Space by Animals and Men*, ed. A. H. Esser, pp. 92–111. New York: Plenum Press.

Williamson, K. (1965). *Fair Isle and Its Birds*. Edinburgh: Oliver and Boyd.

Wilson, E. O. (1975). *Sociobiology*. Massachusetts: Belknap Press.

Witherby, H. F., Jourdain, F. C. R., Ticehurst, N. F. & Tucker, B. W. (1941). *Handbook of British Birds*, vol. 5. London: H. F. and G. Witherby.

Wynne-Edwards, V. C. (1962). *Animal Dispersion in Relation to Social Behaviour*. Edinburgh: Oliver and Boyd.

Yates, F. (1934). The analysis of multiple classifications with unequal numbers in the different classes. *Journal of the American Statistical Association*, **29**, 51–66.

Yeates, G. K. (1948). *Bird Haunts in Northern Britain*. London: Faber.

Index